NON AD PERNI- CIEM

Chaotic Transitions in Deterministic and Stochastic Dynamical Systems

PRINCETON SERIES IN APPLIED MATHEMATICS

The Princeton Series in Applied Mathematics publishes high quality advanced texts and monographs in all areas of applied mathematics. Books include those of a theoretical and general nature as well as those dealing with the mathematics of specific applications areas and real-world situations.

Chaotic Transitions in Deterministic and Stochastic Dynamical Systems

Applications of Melnikov Processes in Engineering, Physics, and Neuroscience

Emil Simiu

PRINCETON UNIVERSITY PRESS
PRINCETON AND OXFORD

Published by Princeton University Press, 41 William Street, Princeton, New Jersey 08540

In the United Kingdom: Princeton University Press, 3 Market Place, Woodstock, Oxfordshire OX20 1SY

Library of Congress Cataloging-in-Publication Data

Simiu, Emil.
Chaotic transitions in deterministic and stochastic dynamical systems: applications of Melnikov processes in engineering, physics, and neuroscience / Emil Simiu.
p. cm. (Princeton series in applied mathematics)
Includes bibliographical references and index.
ISBN 0-691-05094-5 (acid-free paper)
1. Differentiable dynamical systems. 2. Chaotic behavior in systems. 3. Stochastic systems. I. Title.

QA614.8 .S55 2002
515′.352—dc 21

2001059163

British Library Cataloging-in-Publication Data is available

This book has been composed in Times Roman and Abadi
Printed on acid-free paper. ∞
www.pupress.princeton.edu

Printed in the United States of America

10 9 8 7 6 5 4 3 2 1

Et ce monde rendait une étrange musique,
Comme l'eau courante et le vent,
Ou le grain qu'un vanneur d'un mouvement rhytmique
Agite et tourne dans son van.

—Charles Baudelaire, *Une charogne* (Les Fleurs du Mal)

Contents

Preface

A wide variety of natural phenomena can be modeled as planar dynamical systems with motions that may exhibit deterministically or stochastically induced transitions, that is, escapes from and captures into distinct regions in which the systems' motions can evolve. For example, the rolling motion of a vessel may escape from a safe region and be captured into a region wherein capsizing occurs. Mathematically, the transitions entail the crossing of barriers between potential wells.

The Melnikov method provides necessary conditions for the occurrence of such transitions, as well as useful insights into how the transitions are affected by the system and excitation characteristics. It provides a unified treatment of deterministic systems, systems with various types of stochastic excitation—additive, state dependent, white, colored, Gaussian, continuously distributed non-Gaussian, or dichotomous—and systems with combinations of deterministic and stochastic excitation. It has wide-ranging application in science and engineering: in this book alone examples of applications range from physics to mechanical engineering, naval architecture, oceanography, nonlinear control, stochastic resonance, and neurophysiology. It is an effective tool for modeling complex dynamical systems. It allows a chaotic dynamics interpretation of stochastically induced transitions. Last but not least, it is elegant and transparent.

This book is designed to introduce the Melnikov method to readers interested primarily in applications. Prerequisites for the development of stochastic Melnikov theory are (1) deterministic Melnikov theory and (2) basic elements of the theory of stochastic processes. Therefore, following an introduction (Chapter 1), Melnikov theory is presented in a deterministic context, first as a tool applicable to motions with transitions in planar systems, without reference to chaotic dynamics (Chapter 2), then in relation to the chaotic nature of such motions (Chapter 3). These chapters are designed for readers with no previous exposure to dynamical systems theory. Chapter 4 presents requisite elements of the theory of stochastic processes. Based on material presented in Chapters 2, 3, and 4, Chapter 5 extends Melnikov theory to motions with transitions in stochastic planar systems. Chapters 2 through 5 form Part 1 of the book and are devoted to fundamentals. Part 2 consists of Chapters 6 through 12 and is devoted to applications.

The material of Chapters 2 to 4 allows readers not familiar with the theory of nonlinear dynamical systems and/or the theory of stochastic processes to acquire the background needed for the applications without having to resort to a mass of specialized texts less focused with respect to the material specifically needed in this book and more elaborate mathematically. The fundamental material presented in Chapter 3 can be omitted on a first reading as it is not used in all applications. It is needed, however, for understanding the chaotic behavior of the systems being studied, in particular systems exhibiting stochastic resonance.

One of the themes emerging from this book is the hitherto virtually unexplored relationship between chaotic and stochastic dynamics. To our knowledge this contains the first published material that deals with this relationship and explains it within the framework of Melnikov theory.

In Part 2 of the book each of the chapters is concerned with a particular type of application and can be read independently of the others. Material covered in the Appendixes may be omitted on a first reading with no significant prejudice to the reader's ability to apply the method's basic results.

The book is to a large extent self-contained, and numerous references are provided for basic material and additional details. Prerequisites for the book are the equivalent of a first-year graduate course in applied mathematics, including systems of linear differential equations. Laborious mathematical derivations were kept to a minimum, with a few exceptions: the derivation of the expression for the Melnikov function (Chapter 2 and Appendix A1), and the material on the Smale horseshoe map and the shift map (Chapter 3), which convincingly reveals the essence and beauty of chaotic dynamics.

I am indebted to Dr. John W. Lyons and Dr. Richard N. Wright of the National Institute of Standards and Technology for their steadfast support of my initial efforts in the field of chaotic dynamics, and to Professor Stephen Wiggins of the California Institute of Technology for his encouragement and helpful advice during the initial phases of my research. Support by the Ocean Engineering Division of the Office of Naval Research (Dr. Steven Ramberg, Dr. Michael Shlesinger, and Dr. Tom Swean) is also acknowledged with thanks. Dr. Michael R. Frey and Dr. Marek Franaszek contributed many original and stimulating ideas to the research covered in this book. Their contributions are gratefully acknowledged, as are those of Dr. Graham R. Cook, Dr. Charles Hagwood, and Dr. Yudaya Sivathanu. I am also grateful to Dr. Kevin J. Coakley, Professor Mircea Grigoriu, Dr. Howland A. Fowler, Dr. Agnessa Kovaleva, Dr. David Sterling, and Professor Timothy M. Whalen, and to the Princeton University Press reviewers, for many valuable comments and criticisms. It is a pleasure to acknowledge the professionalism and helpfulness of Mr. Trevor Lipscombe, Mr. David Ireland, and Ms. Vickie Kearn (editors), Ms. Anne Reifsnyder (production editor), and Ms. Jennifer Slater, of Princeton University Press.

I dedicate this book to Devra.

Credits

Figures 2.3, 2.4, 2.5, and 3.9, after D. K. Arrowsmith and C. M. Place, *An Introduction to Dynamical Systems* (1990), Cambridge University Press. Reprinted with the permission of Cambridge University Press.

Figure 2.11, after *Nonlinearity*, Vol. 4, D. Beigie, A. Leonard, and S. Wiggins, "Chaotic transport in the homoclinic and heteroclinic tangle region of quasiperiodically forced two-dimensional dynamical systems," 775–819 (1981). Reproduced with the permission of S. Wiggins.

Figure 2.14, after *Journal of Applied Mechanics*, Vol. 55, S. Wiggins and S. Shaw, "Chaos and 3-D horseshoes in slowly varying oscillators," 959–968 (1988). Reproduced with the permission of the American Society of Mechanical Engineers.

Figure 3.4, after *Nonlinear Dynamics and Chaos*, J.M.T. Thompson and H. B. Stewart, John Wiley and Sons Ltd. (1986). Reproduced with the permission of John Wiley and Sons Limited.

Figures 3.10a, 3.11a, and 3.12a,b, from J. Guckenheimer and P. Holmes, *Nonlinear Oscillations, Dynamical Systems, and Bifurcations of Vector Fields* (1986), pp. 89–90, Figs. 2.2.7b,c and 2.2.8a,b, copyright 1983 by Springer-Verlag New York, Inc. Reproduced with the permission of Springer-Verlag and P. Holmes.

Figures 6.3 and 6.4, from *Proceedings of the Royal Society of London*, Vol. A446, S.-R. Hsieh, A. W. Troesch, S. W. Shaw, "A non-linear probabilistic method for predicting vessel capsizing in random beam seas," 195–211 (1994). Reproduced with the permission of the Royal Society and S. W. Shaw.

Figures 11.1 and 11.2, after *Journal of Fluid Mechanics*, Vol. 226 (1991): 511–547. J. S. Allen, R. M. Samuelson, and P. A. Neuberger, "Chaos in model of forced quasi-geostrophic flow over topography: An application of the Melnikov method." Reprinted with the permission of Cambridge University Press.

Figures 12.2 and 12.3, after *Journal of Neurophysiology*, Vol. 30, J. R. Rose, J. F. Brugge, D. J. Anderson, and J. F. Hind, "Phase-locked Response to Low-frequency Tones in Single Auditory Nerve Fibers of the Squirrel Monkey," 769–783 (1967), and Vol. 36, M. A. Ruggero, "Response to Noise of Auditory Nerve Fibers of the Squirrel Monkey," 569–587 (1973), respectively. Reproduced with the permission of the American Physiological Society.

Figure 12.4, reprinted from *Neuroscience*, Vol. 13, I. J. Hochmair-Desoyer, E. S. Hochmair, H. Motz, and F. Rattay, "A Model for the Electrostimulation of the Nervus Acusticus," 553–562 (1984). Reproduced with permission from Elsevier Science.

Figure 12.5, after *Journal of Statistical Physics*, Vol. 70, A. Longtin, "Stochastic Resonance in Neuron Models," 309–327 (1993), Plenum Publishers. Reproduced with the permission of Kluwer Academic/Plenum Publishers and A. Longtin.

Chaotic Transitions in Deterministic and Stochastic Dynamical Systems

Chapter One

Introduction

This work has two main objectives: (1) to present the Melnikov method as a unified theoretical framework for the study of transitions and chaos in a wide class of deterministic and stochastic nonlinear planar systems, and (2) to demonstrate the method's usefulness in applications, particularly for stochastic systems. Our interest in the Melnikov method is motivated by its capability to provide criteria and information on the occurrence of transitions and chaotic behavior in a wide variety of systems in engineering, physics, and the life sciences.

To illustrate the type of problem to which the Melnikov method is applicable we consider a celebrated experiment on a system known as the magnetoelastic beam. The experiment demonstrates the remarkable type of dynamic behavior called *deterministic chaos* (Moon and Holmes, 1979). The system consists of (a) a rigid frame fixed onto a shaking table that may undergo periodic horizontal motions, (b) a beam with a vertical undeformed axis, an upper end fixed onto the frame, and a free lower end, and (c) two identical magnets equidistant from the undeformed position of the beam (Fig. 1.1). The beam experiences nonlinear displacement-dependent forces induced by the magnets, linear restoring forces due to its elasticity, dissipative forces due to its internal friction, the viscosity of the surrounding air, and magnetic damping, and periodic excitation forces due to the horizontal motion of the shaking table. Neither the system properties nor the forces acting on the beam vary randomly with time: the system is fully deterministic.

In the absence of excitation, and depending upon the initial conditions, the beam settles on one of two possible stable equilibria, that is, with the beam's tip closer to the right magnet or closer to the left magnet. The beam also has an unstable equilibrium position—its vertical undeformed axis.

If the excitation is periodic, three distinct types of steady-state dynamic behavior can occur:

1. For sufficiently small excitations, depending again upon the initial conditions, the beam moves periodically about one of its two stable equilibria. The periodic motion is confined to a half-plane bounded by the beam's unstable equilibrium position (Fig. 1.2(a)); in this type of motion there can be no escape from that half-plane.

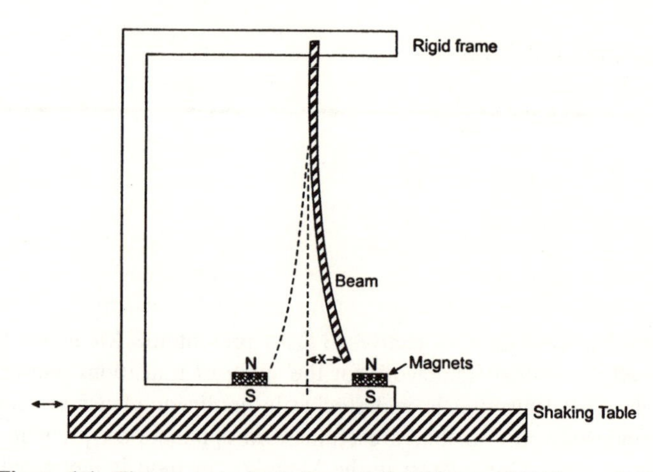

Figure 1.1. The magnetoelastic beam (after Moon and Holmes, 1979).

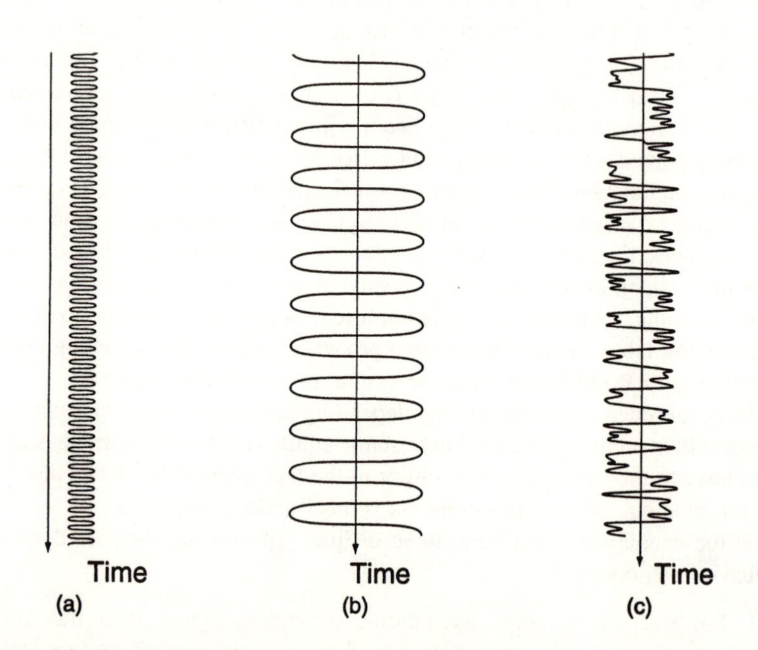

Figure 1.2. Types of steady-state dynamic behavior that may be observed in the periodically excited magnetoelastic beam: (a) periodic motion confined to one of the half-planes bounded by the beam's unstable equilibrium position; (b) periodic motion visiting both half-planes; (c) irregular motion with transitions.

2. For sufficiently large excitations the motion is periodic about—and crosses periodically—the unstable equilibrium position (Fig. 1.2(b)).
3. For intermediate excitation amplitudes, and for restricted sets of initial conditions and excitation frequencies, the steady-state motion is *irregular*, even though the system is fully deterministic; hence the term deterministic chaos. The motion evolves about one of the three equilibria, then it undergoes successive *transitions*, that is, it changes successively to motion about one of the other two equilibria (Fig. 1.2(c)). Transitions in such irregular, deterministic motion are referred to as *chaotic*. A transition away from motion in a half-plane bounded by the beam's unstable equilibrium position is called an *escape*. A transition to motion occurring within such a half-plane is called a *capture*. A succession of escapes and captures is referred to as *hopping*.

The system just described may be modeled as a *dynamical system*—a system that evolves in time in accordance with a specified mathematical expression. In this book we are primarily concerned with dynamical systems capable of exhibiting all three types of behavior illustrated in Fig. 1.2. One basic feature of such systems is that they are *multistable*, meaning that their unforced counterparts have at least two stable equilibria (the term applied to the case of two stable equilibria is *bistable*). In the particular case of mechanical systems, the dynamic behavior is modeled by nonlinear differential equations expressing a relationship among terms that represent

- inertial forces
- dissipative forces
- potential forces, that is, forces derived from a potential function and dependent solely upon displacements; for Fig. 1.1 these forces are due to the magnets and the elasticity of the beam
- excitation forces dependent explicitly on time

Similar terms occur in equations modeling other types of dynamical system, for example, electrical, thermal, or chemical systems.

For a large number of systems arising in engineering or physics safe operation requires that steady-state motions occur within a restricted region, called a *safe region* (in Fig. 1.2(a), the displacement coordinates in the restricted regions are bounded by the vertical line that coincides with the axis of the undeformed beam); transitions to motions visiting another region are undesirable. However, for some systems (e.g., systems that enhance heat transfer, and neurological systems whose activity entails escapes or, in neurological terminology, firings) the occurrence of such transitions is a functional requirement.

Although we will also examine systems with a slowly varying third variable, our main focus will be on planar systems, that is, continuous systems with two time-dependent variables, for example, displacement and velocity. For planar systems subjected to periodic excitation an analytical condition that guarantees the nonoccurrence of transitions was derived in a seminal paper by Melnikov (1963). That condition involves a function—the *Melnikov function*—consisting of a sum of terms related, through the system's potential, to the system dissipation and excitation. The *Melnikov condition for nonoccurrence of transitions* states that if the Melnikov function has no zeros or at most a double zero, then transitions cannot occur. The counterpart of that condition, where the Melnikov function has simple zeros, is referred to as *the Melnikov necessary condition for the occurrence of transitions*. Melnikov theory requires that the perturbation—that is, the excitation and dissipation—be sufficiently small. Numerical calculations show, however, that the theory is useful for perturbations large enough to be of interest in a wide variety of applications.

Melnikov was motivated by problems arising in nuclear physics and engineering (e.g., the theory of particle accelerators, controlled thermonuclear reactions). However, his condition has subsequently found application in a wide variety of disciplines. Examples of applications, some of which are reviewed in subsequent chapters, include

- electronic devices such as the Josephson junction (Genchev, Ivanov, and Todorov, 1983)—see Section 2.4
- ships—for which the roll angle must not exceed the value beyond which capsizing occurs (Falzarano, Shaw, and Troesch, 1992)
- transversely excited buckled columns, which may experience snap-through, that is, motion across the column's undeformed axis (Holmes and Marsden, 1981)
- counter-rotating eccentric cylinders with near-parallel lubricating flow, in which heat transfer is enhanced by chaotic behavior (Ghosh, Chang, and Sen, 1992)
- ocean flows induced by fluctuating wind along a continental shelf with corrugated floor (Allen, Samelson, and Newberger, 1991)
- systems for which control forces are used to suppress or modify the onset of chaotic motions (Lima and Pettini, 1990; Cicogna and Fronzoni, 1993)
- modulations for forcing a phased-locked loop FM demodulator circuit into chaotic behavior (Booker and Smith, 1999, Bishop and Thompson, 1999)

The Melnikov method consists of the application, in various forms, of the necessary Melnikov condition for the occurrence of transitions. So far we have mentioned Melnikov's original method which, as indicated earlier, was restricted to periodically excited planar systems. The extension of the Melnikov method to planar systems with *nonperiodic* excitation is of interest.

For deterministic systems with *quasiperiodic* excitation (i.e., excitation consisting of a sum of harmonics with, in general, incommensurate frequencies) the extension was developed by Wiggins (1988).

The extension to *stochastic* systems was introduced by Frey and Simiu (1993). It is referred to as the *stochastic Melnikov method*—Melnikov's method applied to stochastic dynamical systems. The stochastic counterpart of the Melnikov function is referred to as a *Melnikov process*.

For systems whose stochastic excitations are bounded the stochastic Melnikov method is a straightforward extension of the original Melnikov method. Systems excited by dichotomous noise, which are of interest in electrical engineering and chemical physics, are one example. However, for systems whose stochastic excitations are modeled by unbounded processes the Melnikov method cannot be used to guarantee the nonoccurrence of escapes. Rather, it can be used to calculate lower bounds for probabilities that transitions will not occur during specified time intervals, or upper bounds for probabilities that motion with transitions can occur.

Melnikov (1963) had noted that, depending upon whether it has or does not have simple zeros, the function he developed can be associated with distinct types of dynamic behavior. However, he obtained his original results too early in the development of the theory of chaotic dynamics to describe one of those types of behavior as deterministic chaos. It has been known for several decades that, for deterministic planar multistable systems, the condition that the Melnikov function have simple zeros is necessary for the occurrence not only of transitions but of chaotic behavior as well (see Chapter 3). In fact, for such systems, motions with transitions occur if and only if the motion is chaotic. However, that motions with transitions occurring in stochastic planar multistable systems are also chaotic has only been established relatively recently (see Chapter 5).

The following feature of the Melnikov method is especially useful in applications. Through the *Melnikov scale factor*—a function of frequency that depends upon the system's potential—the Melnikov method provides useful information on the relative degree to which the various frequency components of the excitation are effective in promoting transitions. We discuss in this book several applications of this feature, including

- Efficient control strategies for reducing the probability of stochastically induced escapes. By using information inherent in the Melnikov scale factor it is possible to devise control systems ensuring that most of the control system's energy is spent on counteracting excitation components that promote escapes effectively. In the absence of such information much or in some instances most of the energy would be wasted on counteracting excitation components that contribute negligibly to the promotion of escapes.

- Models of stochastic systems whose observed motions exhibit chaotic transitions. An interesting neurophysiological application is the modeling of the auditory nerve fiber. Melnikov-based models yield good agreement with experimental observations of behavior induced by periodic, quasiperiodic, and broadband white noise excitation, including observations that the classical Fitzhugh-Nagumo model appears to reproduce incorrectly. They also reveal that the dynamics of the auditory nerve fiber is chaotic.
- Stochastic resonance. This designates the phenomenon wherein, for signals contaminated by noise, the signal-to-noise ratio can under certain conditions be improved by the apparently paradoxical means of *increasing* the noise. An interpretation of stochastic resonance based on stochastic Melnikov theory shows that this phenomenon is chaotic, and explains in a transparent fashion the role of the frequency distribution of the noise in the enhancement of the signal-to-noise ratio. That interpretation also leads naturally to extending the framework of stochastic resonance, which can be shown to include the enhancement of the signal-to-noise ratio not only through the addition of noise excitation, but also through the addition of a harmonic excitation, a means that under certain conditions can be considerably more effective.

Following this Introduction, the book is divided into two parts. Part 1, containing Chapters 2 through 5 is devoted to fundamentals. Part 2, containing Chapters 6 through 12, includes applications. In Chapter 2 we introduce basic elements of the theory of nonlinear dynamical systems needed for the development of the Melnikov method. We then present the derivation of the Melnikov function and the Melnikov necessary condition for the occurrence of transitions. For a particular type of system we interpret this condition from an energy viewpoint. We discuss two useful tools that simplify the study and facilitate the diagnosis of motions in continuous dynamical systems: the Poincaré section, which is applicable to periodically excited systems, and the phase space slice, which is the counterpart of the Poincaré section for quasiperiodically excited systems. We introduce the notion of transport associated with transitions, and a useful measure of the transport, the Melnikov functional known as the phase space flux factor. Finally, for the case of slowly varying planar systems, we present the derivation of the Melnikov function and the associated necessary condition for the occurrence of transitions, and note a shortcoming of the Melnikov approach with respect to the case of planar systems with no slowly varying parameter.

In Chapter 3 we establish that motions with transitions—motions for whose occurrence it is necessary that the Melnikov function have simple zeros—are chaotic. As a first step we provide introductory definitions and illustrations of Cantor sets and fractal dimensions. We then study Smale's famous horseshoe map and the shift map, and show the equivalence of the dynamics induced by these two maps. We introduce simple symbolic dynamics techniques and apply them to the shift map to prove that its behavior is chaotic. This establishes that the equivalent dynamics induced by the Smale horseshoe map is chaotic as well. We define chaos mathematically and show that its properties include sensitivity to initial condition and fractal dimensions of geometrical structures associated with the motion, a property that is readily identifiable in certain types of numerical simulations and can therefore serve as a helpful indicator of deterministic chaos. We then state the Smale-Birkhoff theorem, which establishes that the dynamics of motions with transitions in planar multistable systems is chaotic by showing that it is equivalent to the chaotic dynamics induced by the Smale horseshoe map. We provide illustrative numerical examples and examples of chaotic motions of an experimental system, the buckled Stoker column.

In Chapter 4 we present simple elements of the theory of stochastic processes needed for the subsequent development of the stochastic Melnikov method. In particular, we define stationary stochastic processes, autocovariance and cross-covariance functions, spectral densities, linear filters, convolutions, and transfer functions. We also derive the relation between the spectral density of the output and the spectral density of the input of a linear filter with known transfer function. (A brief review of basic elements of probability theory is presented in Appendix A5.)

In Chapter 5 we develop the stochastic Melnikov method. We show that, like deterministic systems, multistable planar systems with stochastic excitation can exhibit motions with transitions that are in practice indistinguishable from deterministic chaotic motions and, like the latter, are sensitive to initial conditions. We refer to such motions as chaotic stochastic motions, as distinct from stochastic motions that do not exhibit properties associated with chaos. The applicability of the Melnikov method to stochastic systems is demonstrated by approximating the stochastic excitation processes by ensembles of sums of periodic terms, thus reducing the core of the stochastic Melnikov method to the problem, studied in Chapters 3 and 4, of the Melnikov method as applied to systems with periodic or quasiperiodic excitation. By using the stochastic Melnikov method we develop criteria that guarantee the nonoccurrence of transitions in systems excited by dichotomous noise, and simple assessments of the influence of the excitation's spectral shape on the system's escape rate.

We now list the applications of the Melnikov method contained in Part 2. Chapter 6 describes an application to a topic in naval architecture: the capsizing problem for a rolling vessel under the action of wave forces. Chapter 7 describes the application of the Melnikov method to open-loop control. Chapter 8 uses the Melnikov method to interpret the stochastic resonance phenomenon in chaotic dynamics terms and presents useful results that follow from this interpretation. Chapter 9 describes the use of the Melnikov method for determining the cutoff point for excitation by experimental noise in a one-dimensional nonlinear system. Chapter 10 presents a study of a mechanical engineering problem: the snap-through of a buckled column excited by a transverse stochastic loading. Chapter 11 describes an application to a topic in oceanography: the problem of transitions for wind-induced along-shore currents over corrugated bottom topography. Chapter 12 describes an application of the Melnikov method in neurophysiology: the modeling of the behavior of the auditory nerve fiber as a chaotic dynamical system.

PART 1
Fundamentals

Chapter Two

Transitions in Deterministic Systems and the Melnikov Function

The main purpose of this chapter is to show that the necessary condition for the occurrence of transitions in a planar multistable deterministic system with dissipation and excitation is that the system's Melnikov function have simple zeros. To define the Melnikov function and derive its expression we must first introduce a number of simple definitions and results from the theory of nonlinear dynamical systems. We then study the particular case of unperturbed systems, that is, of integrable excitation- and dissipation-free systems. This leads us to the introduction of two special sets, the stable manifold and the unstable manifold, which will be seen to play a fundamental role in the system dynamics. We show that, for unperturbed systems, these manifolds are impermeable in the sense that their presence precludes the occurrence of transitions. Next, we consider the same systems under perturbation by dissipative and forcing terms, and the counterparts in the perturbed systems of the unperturbed systems' stable and unstable manifolds. We can then define the Melnikov function, whose behavior determines the behavior of the perturbed systems' stable and unstable manifolds, which in turn controls the occurrence of transitions. In particular, we can explain in geometric terms the mechanism by which transitions can occur if the Melnikov function has simple zeros. We focus on systems of two ordinary differential equations with dependent variables x_1, x_2, but also consider systems of three equations in which the third variable varies slowly with time.

We now describe the material contained in each of this chapter's sections. In Section 2.1 we are primarily concerned with integrable planar systems. Motions in such systems are organized in relation to the systems' fixed points. In preparation for our geometrical approach we review methods for obtaining fixed points and determining their stability. We illustrate this material for a specific case, the standard Duffing–Holmes equation. For use in fundamental developments taken up in later sections and chapters, we define maps and discuss their fixed and periodic points.

For unperturbed integrable multistable planar systems we define in Section 2.2 two special (nongeneric) types of orbit: homoclinic orbits and heteroclinic orbits. The nongeneric orbits asymptotically emerge from and approach the systems' unstable fixed points. In the plane $\{x_1, x_2\}$ they consist of two coinciding one-dimensional sets, the stable manifold and the unstable manifold. The geometric representation of these sets and of the unperturbed systems' generic orbits leads to a simple explanation of the impossibility of transitions. We then provide illustrative examples of systems with homoclinic and with heteroclinic orbits.

In Section 2.3, in preparation for our derivation of the Melnikov function, we represent the integrable systems' stable and unstable manifolds as coinciding two-dimensional sets in the three-dimensional Cartesian space $\{x_1, x_2, t\}$. In keeping with our geometric approach, we examine the transformation of the two-dimensional stable and unstable manifolds when the integrable systems are perturbed by sufficiently small dissipative and excitation terms, that is, when the systems under consideration are *near integrable*. While the perturbed systems still have a stable manifold and an unstable manfold, these two manifolds no longer coincide, as was the case for the perturbed system, but rather are separated by a distance that depends upon location along the unperturbed homoclinic orbit and is referred to as the *Melnikov distance*.

In Section 2.4 we calculate the *Melnikov function*, which to first order is proportional to the Melnikov distance. In Section 2.5 we show that the Melnikov function of a planar dynamical system with harmonic excitation can be interpreted as the output of a linear *filter*, with frequency equal to the excitation frequency and amplitude proportional to a function of frequency called the *Melnikov scale factor*. We give examples of Melnikov functions for systems whose unperturbed counterparts have homoclinic orbits and heteroclinic orbits. We also obtain the expression for the Melnikov function for systems with quasiperiodic excitation. This expression is used in Chapter 5 for the development of the stochastic Melnikov method. In Section 2.6 we show that if the Melnikov function has simple zeros, then the stable and unstable manifolds of the perturbed systems intersect transversely. For a a particular type of system we interpret this condition from an energy viewpoint.

In Section 2.7 we show that the perturbed systems' intersecting stable and unstable manifolds have a special geometric structure called a *homoclinic tangle*. We study that structure by considering intersections of the manifolds with planes of section normal to the time axis in the three-dimensional space $\{x_1, x_2, t\}$. For systems with periodic excitation, planes of section separated by a time interval equal to the excitation period are called *Poincaré sections*. The latter allow a useful simplification: instead of following the evolution of the motion by considering the continuous differential system, we do so by considering a system with discrete time steps equal to the excitation period—a *Poincaré map*. The *phase space slice* is the counterpart of a Poincaré section for quasiperiodically excited systems. Homoclinic tangles in Poincaré maps and phase space slices allow a clear understanding of the mechanism by which the perturbed systems can experience motions with transitions—motions that are not possible in the systems' integrable counterparts. We also define the *phase space flux factor*, a measure of the frequency of occurrence of transitions that is shown in later chapters to be useful in applications.

In Section 2.8 we consider the case of *slowly varying planar systems*, that is, systems defined by three ordinary differential equations in the time-dependent variables x_1, x_2, x_3, one of which varies slowly with time. We present the expression for the Melnikov function and the associated condition for the occurrence of transitions, and note a shortcoming of the Melnikov approach with respect to the case of planar systems with no slowly varying parameter.

Consideration of the issue of chaos is deferred until Chapter 3, where it is shown that, for planar systems possessing a Melnikov function, deterministic motions with transitions are chaotic.

2.1 FLOWS AND FIXED POINTS. INTEGRABLE SYSTEMS. MAPS: FIXED AND PERIODIC POINTS

Basic elements of the theory of nonlinear dynamical systems are a prerequisite for the subsequent development of Melnikov theory. Sections 2.1.1 and 2.1.2 present definitions and procedures pertaining, respectively, to systems with continuous time variable (flows) and systems with discrete, integer-valued time variable (maps).

2.1.1 Flows. Fixed Points and Their Stability. Integrable Systems

We consider a system of first-order ordinary differential equations with continuous time variable

$$\dot{\mathbf{x}} = \mathbf{f}(\mathbf{x}) \tag{2.1.1a}$$

where $\mathbf{x} = \mathbf{x}(t) = (x_1, x_2, \ldots, x_n)^{\mathrm{T}} \in \mathbf{R}^n$ is a vector function of the independent variable t; $\mathbf{f} = (f_1, f_2, \ldots, f_n)^T$ is a smooth vector function referred to as the *vector field*[1] and defined on some subset $E \subseteq \mathbf{R}^n$, and the overdot indicates differentiation with respect to the time t. The vector field $\mathbf{f}$ is said to generate a flow $\phi_t\colon E \to \mathbf{R}^n$, where $\phi_t(\mathbf{x}) = \phi(\mathbf{x}, t)$ is a smooth function defined for all $\mathbf{x}$ in E and t in an interval $I \subseteq \mathbf{R}$. A solution of Eq. 2.1.1a corresponding to the initial conditions $\mathbf{x}(t_0) = \mathbf{x}_0 \in E$ can be written as $\mathbf{x}(\mathbf{x}_0, t_0; t)$. An alternative notation for such a solution is $\phi(\mathbf{x}_0, t)$ such that $\phi(\mathbf{x}_0, t_0) = \mathbf{x}_0$. The flow can be viewed as the set of all solutions

[1]Bold regular letters used in equations indicate vectors. The transpose of the row $(f_1, f_2, \ldots, f_n)$, denoted by $(f_1, f_2, \ldots, f_n)^{\mathrm{T}}$, is a column vector with components $f_1, f_2, \ldots, f_n$.

of Eq. 2.1.1a with initial conditions in the set $E \in \mathbf{R}^n$ on which the function $\mathbf{f}$ is defined.

It is assumed that $\mathbf{f}$ is C^1 in an open subset E of $\mathbf{R}^n$ containing $\mathbf{x}_0$.[2] Then there exists an $a > 0$ such that Eq. 2.1.1a with the initial condition $\mathbf{x}(t_0) = \mathbf{x}_0$ has a unique solution $\mathbf{x}(t)$ on the time interval $[t_0 - a, t_0 + a]$, where a is a constant.

In scalar form, Eq. 2.1.1a is written

$$\begin{aligned} \dot{x}_1 &= f_1(x_1, x_2, \ldots, x_n), \\ \dot{x}_2 &= f_2(x_1, x_2, \ldots, x_n), \\ &\vdots \\ \dot{x}_n &= f_n(x_1, x_2, \ldots, x_n). \end{aligned} \tag{2.1.1b}$$

If in the expression of the system of differential equations the time variable t does not appear explicitly—as is the case in (2.1.1)—the flow is called *autonomous*. The space of the variables $x_1, x_2, \ldots, x_n$ is called the *phase space* or, for $n = 2$, the *phase plane*. The representation of the motion in the phase plane is referred to as a *phase plane diagram.*

The definitions and procedures that follow will be illustrated in detail in Example 2.1.1 below.

A *fixed point* (also called a zero, an equilibrium, a stationary solution, a singular point, or a critical point) is a solution of Eq. 2.1.1 such that, if $x_1(t_0) = x_{10}$, $x_2(t_0) = x_{20}$, $\ldots$, $x_n(t_0) = x_{n0}$, then $x_1(t) = x_{10}$, $x_2(t) = x_{20}$, $\ldots$, $x_n(t) = x_{n0}$ for all t. The coordinates of a fixed point are obtained by solving for $x_{10}, x_{20}, \ldots, x_{n0}$ the equations

$$f_i(x_{10}, x_{20}, \ldots, x_{n0}) = 0 \ (i = 1, 2, \ldots, n),$$

which, by virtue of Eqs. 2.1.1b, state that the rates of change of the coordinates $x_1, x_2, \ldots, x_n$ vanish at the fixed point $(x_{10}, x_{20}, \ldots, x_{n0})$.

Consider solutions of (2.1.1b) that may be written as

$$x_i(t) = x_{i0} + \delta x_i(t) \ (i = 1, 2, \ldots, n), \tag{2.1.1c}$$

where $\delta x_i(t)$ are small deviations from the fixed point x_{i0} $(i = 1, 2, \ldots, n)$. The equation obtained by substituting (2.1.1c) into (2.1.1b) and neglecting all nonlinear terms in $\delta x_i(t)$ is called the *first variational equation*

$$d[\delta x_i(t)]/dt = \sum_{j=1}^{n} a_{ij} \delta x_j(t) \ (i = 1, 2, \ldots, n), \tag{2.1.2}$$

[2]A function is C^r in an interval if its derivative of rth order exists and is continuous throughout that interval.

where the coefficients $a_{ij} = \partial f_i/\partial x_j$ are evaluated at the fixed point. In the linear approximation, Eq. 2.1.2 provides information on whether solutions near the fixed point stay nearby, that is, information on the *linear stability* of the fixed point.

The character of the solutions of the linear system (2.1.2) depends upon the eigenvalues of the matrix $\{a_{ij}\}$ evaluated at the fixed point. If all the eigenvalues have nonzero real parts the corresponding fixed point is referred to as *hyperbolic*. If, for a hyperbolic fixed point, all the eigenvalues have negative real parts, all nearby solutions of (2.1.1) converge to the fixed point as $t \to \infty$; the fixed point is then called a *stable node* or a *sink*. If all the eigenvalues have positive real parts, the hyperbolic fixed point is called an *unstable node* or a *source*. If some but not all of the eigenvalues have negative real parts while the rest have positive real parts, the hyperbolic fixed point is called a *saddle point*. If the eigenvalues are purely imaginary and nonzero, the nonhyperbolic fixed point is called a *center*.

We now consider the particular case of planar flows (i.e., $n = 2$), under the assumption that a C^2 function $H(x_1, x_2)$ exists such that

$$f_1 = \partial H/\partial x_2, \quad f_2 = -\partial H/\partial x_1. \tag{2.1.3a,b}$$

The function $H(x_1, x_2)$ is referred to as a *Hamiltonian*. The system 2.1.1 then becomes

$$\dot{x}_1 = \partial H/\partial x_2, \tag{2.1.4a}$$

$$\dot{x}_2 = -\partial H/\partial x_1. \tag{2.1.4b}$$

The system 2.1.4 belongs to the class of systems referred to as *Hamiltonian systems*. A fixed point (x_{10}, x_{20}) of (2.1.4) is defined as the solution of the system

$$\partial H/\partial x_2 = -\partial H/\partial x_1 = 0.$$

The matrix of the coefficients a_{ij} of the system's first variational equation is

$$\begin{pmatrix} \partial^2 H/\partial x_2 \partial x_1 & \partial^2 H/\partial x_2^2 \\ -\partial^2 H/\partial x_1^2 & -\partial^2 H/\partial x_1 \partial x_2 \end{pmatrix} = \begin{pmatrix} a & b \\ c & -a \end{pmatrix},$$

where the elements of the matrix are evaluated at the fixed point. The eigenvalues of the matrix are obtained from the characteristic equation $(a - \lambda)(-a - \lambda) - bc = 0$, which yields $\lambda_{1,2} = \pm(a^2 + bc)^{1/2}$. If $a^2 + bc > 0$, the fixed point is hyperbolic, and its eigenvalues are real and of opposite signs; the fixed point is a saddle point. If $a^2 + bc < 0$, then the eigenvalues are purely imaginary, and the fixed point is a center.

Since $dH/dt = (\partial H/\partial x_1)\dot{x}_1 + (\partial H/\partial x_2)\dot{x}_2$ it follows immediately from Eqs. 2.1.4a,b that $dH/dt = 0$, or $H(x_1, x_2) = \text{const}$. In particular, the system

$$f_1 = x_2, \quad f_2 = -V'(x_1) \tag{2.1.4c,d}$$

is Hamiltonian, with the Hamiltonian function $H = x_2^2/2 + V(x_1)$. If Eqs. 2.1.4a–d represent a mechanical system, H represents the system's total energy, that is, the sum of the kinetic and potential energy. Equations 2.1.4a–d can be written in the form

$$\ddot{x}_1 = -V'(x_1). \tag{2.1.5}$$

For the particular case $H(x_1, x_2) = x_2^2/2 + V(x_1)$ the integration of Eqs. 2.1.4a,b can be performed by dividing through Eq. 2.1.4b by Eq. 2.1.4a and separating the variables x_1 and x_2. A first quadrature yields the equation

$$x_2 = \pm[-2V(x_1) + C_1]^{1/2}. \tag{2.1.6}$$

A second quadrature is performed after separating the variables x_1, t in Eq. 2.1.6, in which it is recalled that $x_2 = dx_1/dt$.

Example 2.1.1 *An integrable dynamical system with two centers and a saddle point: the standard Duffing–Holmes oscillator*

To illustrate the definitions and procedures discussed in this section we present in some detail their application to the system 2.1.4a–d, in which the potential function is

$$V(x_1) = -x_1^2/2 + x_1^4/4. \tag{2.1.7}$$

This system is known as the standard *Duffing–Holmes oscillator.*

The potential $V(x_1)$ (Eq. 2.1.7) is depicted in Fig. 2.1a. It is referred to as a double-well potential, its two wells being separated by a potential barrier. A unit gravitational force **F** acting on a particle that slides along the curve $V(x_1)$ may be resolved into a component **N** normal to that curve, which has no effect on the particle motion, and a horizontal component **H**, which drives the particle's motion (Fig. 2.1c). Since the magnitude of **F** is $F = 1$, the magnitude of **H** is $H = H/F = \tan\theta = V'(x_1)$. **H** and the axis Ox_1 have opposite directions; hence the negative sign in the right-hand side of Eq. 2.1.5. The same result is obtained if the balance of the forces **F**, **N**, **H** is considered for a point of the curve $V(x_1)$ at which the slope is negative.

We now seek the fixed points of the system 2.1.4a–d, 2.1.7. Setting $f_1 = f_2 = 0$, we obtain the fixed points (0, 0), $(-1, 0)$, and (1, 0), denoted O, C, and C', respectively. They are represented in the phase plane $\{x_1, x_2\}$

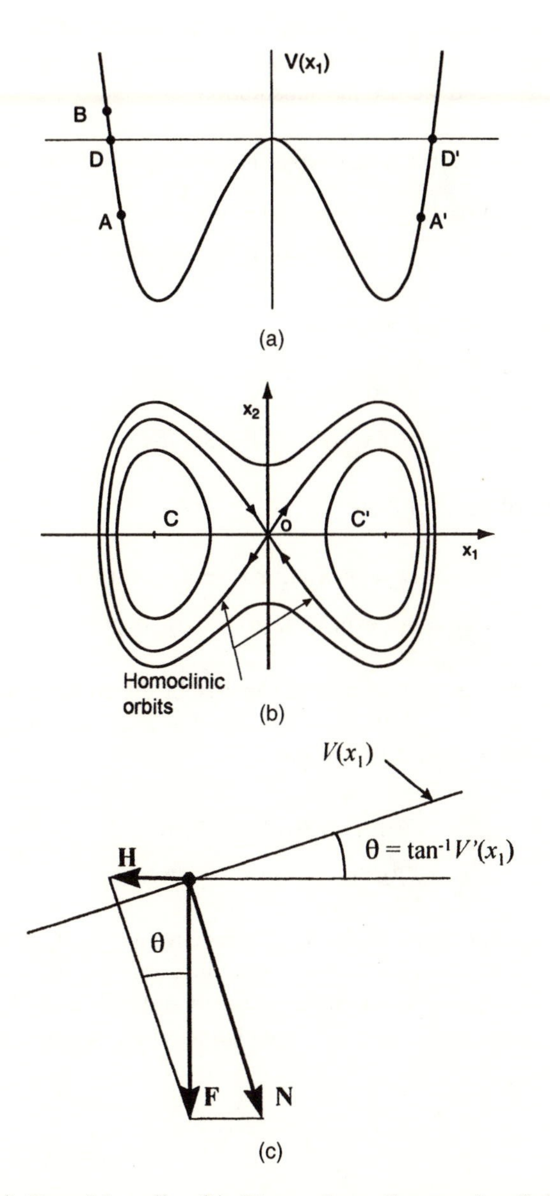

Figure 2.1. (a) Potential wells. (b) Phase plane diagram for Duffing–Holmes equation. The orbits that emerge from and converge to saddle point O are the homoclinic orbits. Periodic orbits confined to a well evolve inside a homoclinic orbit around one of the centers C, C'; those that cross the potential barrier evolve outside the homoclinic orbits. The homoclinic orbits are separatrices between these two types of motion. (c) Balance of forces on a material point moving without excitation and friction on a potential curve.

(Fig. 2.1b). We now examine in some detail the stability of these fixed points by considering the respective variational equations.

If in Eqs. 2.1.4a–d we use the substitution

$$x_i(t) = x_{i0} + \delta x_i(t)\ (i = 1, 2),$$
$$f_i(x_1, x_2) = f_i(x_{10}, x_{20}) + (\partial f_i/\partial x_1)|_{x_1=x_{10}}\delta x_1 + (\partial f_i/\partial x_2)|_{x_2=x_{20}}\delta x_2\ (i = 1, 2),$$

we obtain the variational equations

$$d(\delta x_1)/dt = \delta x_2, \tag{2.1.8a}$$
$$d(\delta x_2)/dt = (1 - 3x_1^2)\delta x_1. \tag{2.1.8b}$$

The solutions of Eqs. 2.1.8 are

$$\delta x_1 = u_{1,s}\exp(\lambda_s t),\ \ \delta x_2 = u_{2,s}\exp(\lambda_s t),$$
$$\delta x_1 = u_{1,u}\exp(\lambda_u t),\ \ \delta x_2 = u_{2,u}\exp(\lambda_u t).$$

Let $A = [a_{ij}]$ $(i, j = 1, 2)$ denote the matrix of the coefficients of $\delta x_1, \delta x_2$ in Eqs. 2.1.8 ($a_{11} = a_{22} = 0, a_{12} = 1, a_{21} = 1 - 3x_1^2$). Substituting the exponential solutions in Eqs. 2.1.8 we get

$$a_{11}u_{1,s} + a_{12}u_{2,s} = \lambda_s u_{1,s}, \tag{2.1.9a}$$
$$a_{21}u_{1,s} + a_{22}u_{2,s} = \lambda_s u_{2,s}, \tag{2.1.9b}$$

with a similar set of equations for λ_u. The eigenvalues λ_s, λ_u at point (x_1, x_2) are obtained from the characteristic equation, that is, from the condition $|A - I\lambda| = 0$, where I is the unit matrix of order 2, and $|A - I\lambda|$ is the determinant of the matrix $A - I\lambda$. The characteristic equation is then

$$\begin{vmatrix} -\lambda & 1 \\ 1 - 3x_1^2 & -\lambda \end{vmatrix} = 0. \tag{2.1.10}$$

Its solutions are

$$\lambda_{u,s} = \pm(1 - 3x_1^2)^{1/2}. \tag{2.1.11}$$

For the fixed point O, $x_1 = 0$ and $\lambda_u = 1$, $\lambda_s = -1$. It follows from this result that the fixed point O is a saddle point.

Using the first equation in Eqs. 2.1.9 and its counterpart for the stable eigenvector, and recalling that $a_{11} = 0$, $a_{12} = 1$, we obtain the slopes of the eigenvectors,

$$u_{2,s}/u_{1,s} = -1, \tag{2.1.12a}$$

$$u_{2,u}/u_{1,u} = 1. \tag{2.1.12b}$$

For the fixed points C and C' both eigenvalues are imaginary. These points are therefore nonhyperbolic and, as mentioned earlier, they are referred to as centers.

2.1.2 Maps. Fixed Points and Their Stability. Periodic Points

A *map* is a dynamical system with discrete, integer-valued time variables. Consider the n-dimensional map

$$x_{ji+1} = f_j(x_{1i}, x_{2i}, \ldots, x_{ni}),\ j = 1, 2, \ldots, n. \tag{2.1.13}$$

For given initial conditions $x_{10}, x_{20}, \ldots, x_{n0}$, the first and $(k+1)$th iterates are, respectively,

$$x_{j1} = f_j(x_{10}, x_{20}, \ldots, x_{n0}), \tag{2.1.14}$$

$$x_{jk+1} = f_j(x_{1k}, x_{2k}, \ldots, x_{nk}) \equiv f_j^{k+1}(x_{10}, x_{20}, \ldots, x_{n0}). \tag{2.1.15}$$

In Eq. 2.1.15 the superscript $k+1$ denotes the order of iteration, rather than being an exponent. The coordinates of a *periodic point of period* m $(m = 1, 2, \ldots)$ with coordinates $x_{10}, x_{20}, \ldots, x_{n0}$ are the solutions of the equations

$$\begin{aligned} x_{jm} &= f_j^m(x_{10}, x_{20}, \ldots, x_{n0}) \\ &= x_{j0},\ j = 1, 2, \ldots, n. \end{aligned} \tag{2.1.16}$$

A *fixed point* is a periodic point of period 1.

Following an approach similar to the approach that led to Eq. 2.1.2, it can be shown that the stability of a periodic point is ascertained in the linear approximation by examining the eigenvalues of the matrix

$$A = \{\partial f_j^m/\partial x_i|_{x=x_0}\}. \tag{2.1.17}$$

A periodic point is hyperbolic if the moduli of all the eigenvalues of A differ from unity. Otherwise the point is nonhyperbolic. A hyperbolic periodic point is of the saddle type if some, but not all, of the eigenvalues of A have moduli greater than 1, and the rest of the eigenvalues have moduli less than 1. If all the eigenvalues have moduli less than 1, then the periodic point is called a stable node or a sink, and solutions near the periodic point converge to the periodic point. If all the moduli are larger than 1 the periodic point is called an unstable node or a source. If the eigenvalues have modulus 1, the nonhyperbolic point is referred to as a center.

Example 2.1.2 *Fixed points of a one-dimensional map and their stability*

We seek the fixed points of the one-dimensional map

$$x_{11} = \tfrac{1}{2}x_{10}(x_{10} - 1).$$

Equation 2.1.16 (with $m = 1$) yields

$$x_{10} = \tfrac{1}{2}x_{10}(x_{10} - 1),$$

that is, the system has two fixed points, with coordinates $x_{10} = 0$ and $x_{10} = 3$. The matrix A (Eq. 2.1.17) has the single element

$$d\left[\tfrac{1}{2}x_1(x_1 - 1)\right]/dx_1\Big|_{x_1 = x_{10}} = \left(x_1 - \tfrac{1}{2}\right)\Big|_{x_1 = x_{10}}.$$

For the fixed point $x_{10} = 0$, the modulus of the eigenvalue is $|\lambda| = \frac{1}{2}$, that is, the fixed point is stable. For the fixed point $x_{10} = 3$, $|\lambda| = 2.5$, that is, the fixed point is unstable.

2.2 HOMOCLINIC AND HETEROCLINIC ORBITS. STABLE AND UNSTABLE MANIFOLDS

In this section we introduce the mathematical objects referred to as stable and unstable manifolds, whose behavior controls the system's transitions.

For the unforced and dissipation-free system with the potential of Fig. 2.1a, let us consider a particle released with zero velocity from a point A (or A') with ordinate $V(x_1) < 0$ (i.e., a point located within a well, below the top of the potential barrier). The particle will move periodically about the center C (or C') located at the bottom of that well. In the phase plane $\{x_1, x_2\}$ the motion is represented by a closed curve surrounding the center C (or C'). If the particle is released with zero velocity from a point B with ordinate $V(x_1) > 0$, it will move periodically in the phase plane about the saddle point O, and during that motion it will visit both wells periodically. Depending upon its initial position on the potential curve and its initial velocity (i.e., depending upon its total energy), a particle may describe any one of an infinity of periodic orbits topologically similar to the orbits depicted in Fig. 2.1b.

Particles whose total energy at the time of release is $x_2^2/2 + V(x_1) = V(0)$ (e.g., particles released with zero velocity from point D or D' in

Fig. 2.1a) describe orbits with infinitely long periods. Such orbits are sometimes referred to as nongeneric. In Fig. 2.1b the motion on these orbits proceeds in forward time away from the saddle point along the eigenvector associated with unstable motion, and then toward the saddle point which it approaches along the eigenvector associated with stable motion. The motion is said to proceed along an *unstable manifold* connecting the saddle point to itself. Alternatively, the motion on a nongeneric orbit may be viewed as proceeding in reverse (negative) time away from the saddle point along the vector associated with stable motion, and then toward the saddle point which it approaches along the eigenvector associated with unstable motion. In this view the motion is said to proceed in reverse time along a *stable manifold* connecting the saddle point to itself. We note that *the stable* and *unstable manifolds coincide.* The nongeneric orbits connecting the saddle point to itself are called *homoclinic.*

Example 2.2.1 *Homoclinic orbits in the standard Duffing–Holmes oscillator*

The equations of the homoclinic orbits of the standard Duffing–Holmes oscillator (Eqs. 2.1.4a–d, 2.1.7) are found by following the procedure outlined in Section 2.1.1 (see Eq. 2.1.6). The homoclinic orbits are defined by the vector

$$\mathbf{x}_h^{\pm}(t) = \{x_{h1}^{\pm}(t), x_{h2}^{\pm}(t)\}^{\mathrm{T}},$$

where the plus and minus signs denote the orbit in the positive and negative x_1 half-planes, respectively. The result of the integration, for which the initial conditions are $x_1(0) = \pm\sqrt{2}$, $x_2(0) = 0$, is

$$\{x_{h1}^{\pm}(t), x_{h2}^{\pm}(t)\} = \{\pm\sqrt{2}\ \mathrm{sech}\ t, \mp\sqrt{2}\ \mathrm{sech}\ t\ \tanh\ t\}. \tag{2.2.1}$$

The initial conditions $x_1(0) = \pm\sqrt{2}$ correspond in Fig. 2.1a to positions D', D.

Example 2.2.2 *Heteroclinic orbits in the rf-driven Josephson junction*

We now give an example of a system with nongeneric orbits that connect distinct saddle points, rather than connecting a saddle point to itself. Such orbits are called *heteroclinic.* Like their homoclinic counterparts, the heteroclinic orbits consist of coinciding stable and unstable manifolds.

We consider the system 2.1.3, 2.1.4 in which

$$V'(x_1) = \alpha \sin x_1. \tag{2.2.2}$$

This system, which represents the equations of motion of a simple frictionless pendulum, is also the unperturbed counterpart of the electronic device known as the radio frequency (rf) model of the Josephson junction (Section 2.4).

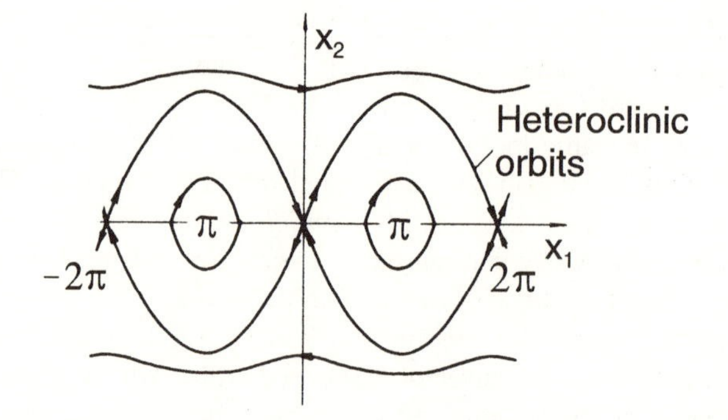

Figure 2.2. Phase plane diagram for simple frictionless pendulum.

The fixed points of the system 2.1.4a–d, 2.2.2 are obtained by setting $x_2 = 0$, $\sin x_1 = 0$. Their coordinates are

$$x_1 = \pm k\pi \, (k = 0, 1, 2, \ldots), \quad x_2 = 0. \tag{2.2.3}$$

The fixed points for which $k = 0, \pm 2, \pm 4, \ldots$ are saddle points. Those for which $k = \pm 1, \pm 3, \pm 5, \ldots$ are centers. Each saddle point is connected to the adjacent saddle points by heteroclinic orbits. The equations of the heteroclinic orbit's half-loop for which $0 \leq x_1 \leq 2\pi$, $x_2 > 0$ (Fig. 2.2) are obtained by the procedure outlined in Section 2.2.1 (see Eq. 2.1.6). The result is

$$\{x_{h1}(t), x_{h2}(t)\} = \{4 \tan^{-1}[\exp(\alpha^{1/2} t)], \, 2\alpha^{1/2} \operatorname{sech}(\alpha^{1/2} t)\}. \tag{2.2.4a,b}$$

Figure 2.2 also shows orbits evolving around the centers inside the heteroclinic orbits, and orbits evolving outside the heteroclinic orbits. For oscillations of the simple pendulum the former are referred to as *librations* (from the Latin "libra," a word known to horoscope readers, meaning balance, pair of scales), while the latter, which involve periodic crossings of the unstable position of equilibrium of the pendulum, are called *rotations*. These two terms are also used to designate periodic motions about a center and a saddle point in other multistable systems, for example, the periodic motions depicted in Figs. 1.2a and 1.2b, respectively.

Motions along homoclinic and heteroclinic orbits are limiting cases for motions occurring at all times inside those orbits as the periods of the motions increase indefinitely. No motion—whether it occurs inside or outside the region of the phase plane enclosed by the coinciding stable and unstable manifolds—can cross the manifolds (see Figs. 2.1b and 2.2). This is implicit in the uniqueness of the system's solutions: if intersections did occur, then the system would have distinct solutions containing the intersection point, one belonging to the manifolds and one crossing the manifolds. The coinciding stable and unstable manifolds—the system's homoclinic or heteroclinic orbits—act as *separatrices*. In unforced, dissipation-free systems the presence of the separatrices precludes the occurrence of motions with transitions such as the motion shown in Fig. 1.2c. That motions with transitions *can* occur in systems with excitation and dissipation will be shown later in this chapter.

In this book we study planar integrable systems perturbed by dissipation and excitation terms. Perturbation renders the systems nonintegrable (for small perturbations the perturbed systems are referred to as *near-integrable*). One notable effect of the perturbation is to cause separation between the stable and unstable manifolds. As will be shown subsequently, that separation provides a simple criterion for determining the regions in parameter space (i.e., the sets of parameter values) for which the perturbed system's motion can cross the separatrix of the unperturbed system. We therefore wish to determine the separation between the stable and unstable manifolds, which is a function of position on the unperturbed manifolds and is referred to as the Melnikov distance. It will be shown, however, that it is convenient to calculate instead a function approximately proportional to that distance, the Melnikov function. Preparatory work for deriving that function is done in the next section.

2.3 STABLE AND UNSTABLE MANIFOLDS IN THE THREE-DIMENSIONAL PHASE SPACE $\{x_1, x_2, t\}$

Stable and unstable manifolds in integrable systems were represented in Section 2.2 in the phase plane $\{x_1, x_2\}$. In nonautonomous perturbed systems the time dimension plays an explicit role in the description of the manifolds and, as will be seen in Section 2.4, in the calculation of their separation. For this reason, and for subsequent developments concerning the case of nonperiodic excitations, it is convenient to represent the manifolds in the three-dimensional space $\{x_1, x_2, t\}$.

2.3.1 Unperturbed Systems

We consider the three-dimensional space with orthogonal coordinate axes Ox_1, Ox_2, Ot (Fig. 2.3). We wish to represent in this space the orbits corresponding in the phase plane $\{x_1, x_2\}$ to homoclinic or heteroclinic orbits.[3] To this end we construct in the three-dimensional space a vertical

[3]Henceforth, unless otherwise indicated, statements pertaining to homoclinic orbits will be valid for heteroclinic orbits as well, and we will therefore omit specific reference to the latter.

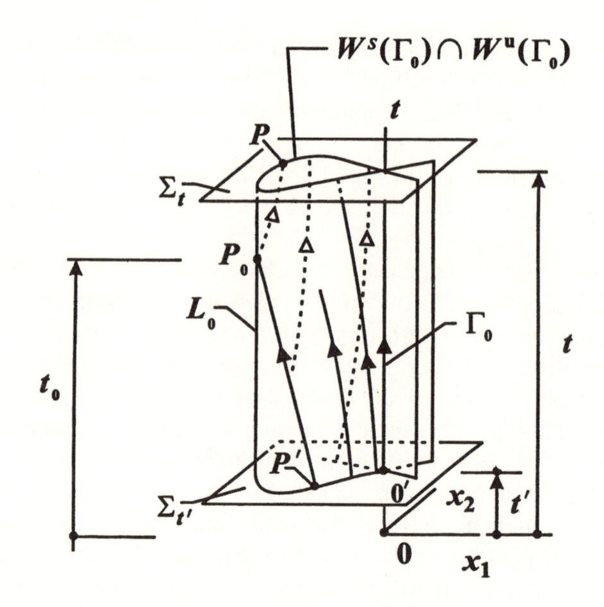

Figure 2.3. Stable and unstable manifolds for unperturbed system. An orbit belonging to the manifolds is defined by cordinate t_0 of the point P_0 at which it is tangent to the fixed reference line L_0. A point P belonging to that orbit is defined by two time coordinates: the coordinate t_0 defining the orbit, and the coordinate $t - t_0$ representing the time of flight between P_0 and P.

cylindrical surface whose directrix consists of the homoclinic orbits of the system being considered. We denote by (x_{h1}, x_{h2}) the coordinates of those orbits in a plane $t =$ const. The intersection of the axis $\Gamma_0 = Ot$ with a plane $t =$ const is a saddle point in that plane.

By construction, the cylindrical surface contains the set of orbits evolving in the space $\{x_1, x_2, t\}$ whose horizontal projections in the phase plane $\{x_1, x_2\}$ are the homoclinic orbits of the system. The set of orbits contained in the cylindrical surface has two subsets: the subset of orbits that emerge asymptotically from the axis $\Gamma_0 \equiv Ot$ and approach it as $t \to \infty$ (these correspond to orbits in the plane $\{x_1, x_2\}$ approaching asymptotically the saddle point of Fig. 2.1b in forward time), and the subset of orbits that approach Γ_0 as $t \to -\infty$ (these correspond to orbits in the plane $\{x_1, x_2\}$ approaching asymptotically the saddle point in reverse time). The two subsets are referred to as the system's unstable manifold $W^u(\Gamma_0)$ and stable manifold $W^s(\Gamma_0)$, respectively. It can be seen in Fig. 2.3 that the stable and unstable manifolds coincide, that is, the cylindrical surface of Fig. 2.3 is the intersection of the stable and unstable manifolds, $W^s(\Gamma_0) \cap W^u(\Gamma_0)$. This is the counterpart in the three-dimensional space of the fact that the homoclinic orbits that approach the saddle point in forward time and those that approach it in reverse time coincide in the phase plane $\{x_1, x_2\}$.

The unstable manifold may be viewed as the union of two sets. The first set belonging to the unstable manifold is a surface contained in an $O(\epsilon)$ neighborhood of Γ_0; it is defined as the *local* unstable manifold of Γ_0. For $\epsilon \to 0$, its intersection with any plane t = const is tangent in that plane to the direction of the eigenvector associated with the positive eigenvalue obtained from the variational equation. The second subset is the *global* unstable manifold of Γ_0, and consists of the set of orbits in forward time whose initial conditions are contained in the local unstable manifold. Similar definitions hold for the local and global stable manifolds. In Appendix A2 it is shown how local manifolds are used for the numerical construction of global manifolds.

The stable and unstable manifolds are *invariant*, that is, they have the property that if a point of an orbit is contained in a manifold, that orbit is contained in the manifold for all time. The set Γ_0 is called *normally hyperbolic*. This means, essentially, that the flow expands or contracts a small vector contained in a local unstable manifold faster if the vector is nearly normal to Γ_0 than if it is nearly parallel to Γ_0.

The coordinate system for the stable and unstable manifolds is represented in Fig. 2.3. We define a fixed vertical reference line L_0 as follows. L_0 is contained in the manifolds and intersects the planes t = const at points with coordinates $\mathbf{x}_h(0)$. An orbit belonging to the manifolds is defined by the time coordinate t_0 of the point P_0 at which the orbit is tangent to L_0. A point P on that orbit is defined by two coordinates: t_0, which defines the entire orbit, and $t - t_0$, which represents the time of flight on the orbit between P_0 and P. The coordinate of the point P in the plane $\{x_1, x_2\}$ is $\mathbf{x}_h(t - t_0)$. In particular, the coordinate of P_0 in that plane is $\mathbf{x}_h(t_0 - t_0) = \mathbf{x}_h(0)$, as was already indicated.

2.3.2 Perturbed Systems

We now consider the perturbed system

$$\dot{\mathbf{x}} = \mathbf{f}(\mathbf{x}) + \epsilon\mathbf{g}(\mathbf{x}, t), \tag{2.3.1}$$

where $\mathbf{x} = (x_1, x_2)^{\mathrm{T}}, \mathbf{f} = (f_1, f_2)^{\mathrm{T}}, \mathbf{g} = (g_1, g_2)^{\mathrm{T}}$. Assume that $0 < \epsilon \ll 1$, $\mathbf{f}$ and $\mathbf{g}$ are bounded and C^r, $r \geq 2$, and the unperturbed system (i.e., Eq. 2.3.1 with $\epsilon = 0$) has a homoclinic orbit $\mathbf{x}_h = [x_{h1}, x_{h2}]^{\mathrm{T}}$ with a saddle point O at the origin.

Under these assumptions, the perturbation transforms the axis Γ_0 into a smooth curve Γ_ϵ whose distance from Γ_0 is $O(\epsilon)$ and depends on t (Fig. 2.4). In addition, the perturbed system has local stable and unstable manifolds whose intersections with a plane of section are C^r-close to the intersections

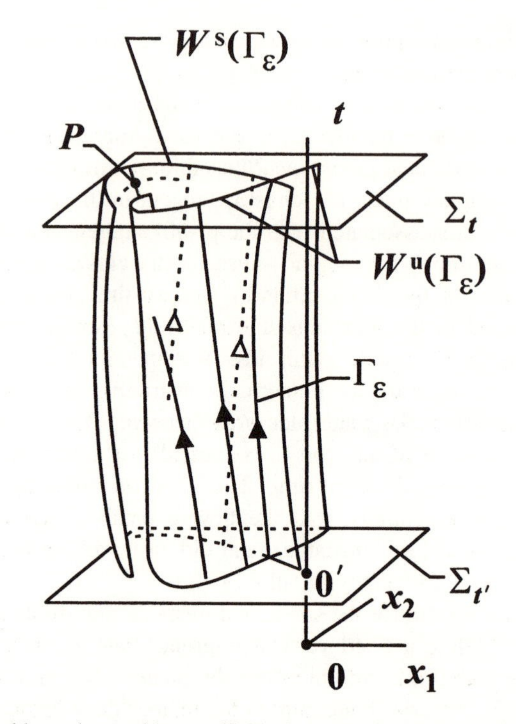

Figure 2.4. Stable and unstable manifolds for perturbed system (after Arrowsmith and Place, 1990). The distance between the separated manifolds at point P of the unperturbed manifold is indicated by the solid line through P. The unperturbed manifold's partial intersection with the plane Σ_t is shown by a dashed line through P.

with that plane of the unperturbed system's local stable and unstable manifolds, respectively, $r \geq 2$.[4] The preceding statements are the substance of the *persistence theorem* (Fenichel, 1971). Like Γ_0, Γ_ϵ is normally hyperbolic. The definitions of the perturbed system's global stable and unstable manifolds, $W^s(\Gamma_\epsilon)$ and $W^u(\Gamma_\epsilon)$, are similar to those applicable to the unperturbed system. As was noted earlier, the perturbed stable and unstable manifolds no longer coincide, as was the case for the unperturbed system. The distance that separates them depends upon position on the unperturbed manifolds and is called the Melnikov distance. It is indicated in Fig. 2.4 at point P of the unperturbed manifolds, whose partial intersection with the plane Σ_t is shown as a dashed line. In the next section we show that, to first order, the Melnikov distance is proportional to the absolute value of the Melnikov function, and we show how the Melnikov function is calculated.

[4]Two elements are C^r-close if they and their first r derivatives are within ϵ as measured in some norm.

2.4 THE MELNIKOV FUNCTION

In this section we derive the expression for the Melnikov function. As will be shown in Section 2.6, the Melnikov function serves as a discriminant that makes it possible to ascertain whether transitions can occur. Further, it will be shown in Section 3.5 that, since transitions imply chaos, the Melnikov function makes it possible to ascertain whether chaotic motion can occur.

In Section 2.4.1 we state the Melnikov theorem, and provide the expression for the Melnikov function. In Section 2.4.2 we define the distance function and show that, to first order, it is proportional to the Melnikov function.

2.4.1 Melnikov Theorem

Consider the system of Eq. 2.3.1,

$$\dot{\mathbf{x}} = \mathbf{f}(\mathbf{x}) + \epsilon \mathbf{g}(\mathbf{x}, t),$$

where $\mathbf{x} = (x_1, x_2)^{\mathrm{T}}$, $\mathbf{f} = (f_1, f_2)^{\mathrm{T}}$, $\mathbf{g} = (g_1, g_2)^{\mathrm{T}}$. If the unperturbed system is Hamiltonian (i.e., $f_1 = \partial H/\partial x_2$, $f_2 = -\partial H/\partial x_1$) and has a homoclinic orbit $\mathbf{x}_h = [x_{h1}, x_{h2}]^{\mathrm{T}}$ with a saddle point O at the origin, $0 < \epsilon \ll 1$, and $\mathbf{f}$ and $\mathbf{g}$ are bounded and C^r, $r \geq 2$, then, to first order, the Melnikov distance is proportional to the absolute value of the Melnikov function

$$M(t_0) = \int_{-\infty}^{\infty} \mathbf{f}[\mathbf{x}_h(\zeta)] \wedge \mathbf{g}[\mathbf{x}_h(\zeta), \zeta + t_0]\, d\zeta, \tag{2.4.1}$$

where the symbol $\wedge$ denotes the wedge product.[5]

2.4.2 The Distance Function and Its First-Order Approximation

Consider the plane $\Sigma_{t'}$ normal to the axis Ot at the ordinate t', its intersection with the unperturbed system's coinciding stable and unstable manifolds (solid lines), and its intersection with the perturbed system's separated stable and unstable manifolds (dashed lines) (Fig. 2.5). We fix the ordinate t'. Point P' is then defined by the coordinates t_0, t' (see Fig. 2.3). Line N is normal at P' to the intersection of the unperturbed manifolds with $\Sigma_{t'}$. We now consider the orbits in the perturbed stable manifold and unstable manifold that intersect N at points A' and A'', respectively.

[5]Recall that the wedge product $\mathbf{a} \wedge \mathbf{b}$ of the vectors $\mathbf{a} = \{a_1, a_2\}^{\mathrm{T}}$ and $\mathbf{b} = \{b_1, b_2\}^{\mathrm{T}}$, with components defined in a Cartesian coordinate system, is the scalar $a_1 b_2 - a_2 b_1 = |\mathbf{a}||\mathbf{b}| \sin \phi$, where ϕ is the angle between $\mathbf{a}$ and $\mathbf{b}$.

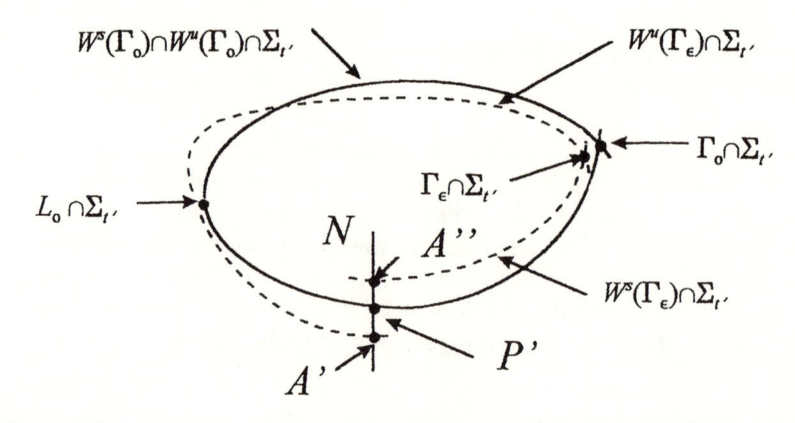

Figure 2.5. Intersections of Γ_0, $W^s(\Gamma_0) \cap W^u(\Gamma_0)$, L_0, Γ_ϵ, $W^s(\Gamma_\epsilon)$, and $W^u(\Gamma_\epsilon)$ with plane $\Sigma_{t'}$. Segment $A'A''$ is the Melnikov distance at the point P' belonging to the unperturbed manifolds (after Arrowsmith and Place, 1990).

These orbits intersect the plane of section with elevation t, denoted Σ_t, at points with coordinates $\mathbf{x}^s(t; t_0, t', \epsilon) = [x_1^s(t; t_0, t', \epsilon), x_2^s(t; t_0, t', \epsilon)]^{\mathrm{T}}$ and $\mathbf{x}^u(t; t_0, t', \epsilon) = [x_1^u(t; t_0, t', \epsilon), x_2^u(t; t_0, t', \epsilon)]^{\mathrm{T}}$, respectively. As indicated earlier, the orbit belonging to the unperturbed manifolds that contains P' intersects Σ_t at a point P with coordinates $\mathbf{x}_h(t - t_0)$. Note also that from Eq. 2.3.1 in which $\epsilon = 0$ it follows that the vector $\mathbf{f}[\mathbf{x}_h(t - t_0)]$ is tangent at P to the intersection of the unperturbed manifolds with Σ_t.

The distance function at elevation t (i.e., in the plane Σ_t) is defined as

$$\Delta_\epsilon(t; t_0, t') = \mathbf{f}[\mathbf{x}_h(t - t_0)] \wedge [\mathbf{x}^u(t; t_0, t', \epsilon) - \mathbf{x}^s(t; t_0, t', \epsilon)]. \tag{2.4.2}$$

Equation 2.4.2 states that the function $\Delta_\epsilon(t; t_0, t')$ is equal to $|\mathbf{f}[\mathbf{x}_h(t - t_0)]|$ times the component of the vector $[\mathbf{x}^u(t; t_0, t', \epsilon) - \mathbf{x}^s(t; t_0, t', \epsilon)]$ along the normal to $\mathbf{f}[\mathbf{x}_h(t - t_0)]$. The absolute value of that component is the Melnikov distance at point P (see Fig. 2.4). In particular, $\Delta_\epsilon(t'; t_0, t')$ is the distance function in the plane $\Sigma_{t'}$ at point P'. Its absolute value is proportional to the Melnikov distance at P', which by definition is equal to the distance between points A' and A'' (see Fig. 2.5).

To obtain an expression that approximates the distance function to first order we proceed as follows. We expand $\mathbf{x}^u(t; t_0, t', \epsilon)$ and $\mathbf{x}^s(t; t_0, t', \epsilon)$ in a power series in ϵ:

$$\mathbf{x}^u(t; t_0, t', \epsilon) = \mathbf{x}_h(t - t_0) + \epsilon \mathbf{x}_1^u(t; t_0, t') + O(\epsilon^2), \tag{2.4.3a}$$

$$\mathbf{x}^s(t; t_0, t', \epsilon) = \mathbf{x}_h(t - t_0) + \epsilon \mathbf{x}_1^s(t; t_0, t') + O(\epsilon^2), \tag{2.4.3b}$$

where $\mathbf{x}_1^u = [x_{11}^u, x_{12}^u]^{\mathrm{T}}$ and $\mathbf{x}_1^s = [x_{11}^s, x_{12}^s]^{\mathrm{T}}$ are first variations with respect to ϵ. Equations 2.4.3a and 2.4.3b are valid uniformly (i.e., for ϵ independent of t) for $t' \le t < \infty$ and $-\infty \le t < t'$, respectively (Guckenheimer and Holmes, 1986, p. 186). Substituting Eqs. 2.4.3 into Eq. 2.4.2 we obtain

$$\Delta_\epsilon(t; t_0, t') = \Delta_\epsilon^u(t; t_0, t') - \Delta_\epsilon^s(t; t_0, t') + O(\epsilon^2), \tag{2.4.4}$$

$$\Delta_\epsilon^u(t; t_0, t') = \mathbf{f}[\mathbf{x}_h(t - t_0)] \wedge \epsilon \mathbf{x}_1^u(t; t_0, t') + O(\epsilon^2), \tag{2.4.5a}$$

$$\Delta_\epsilon^s(t; t_0, t') = \mathbf{f}[\mathbf{x}_h(t - t_0)] \wedge \epsilon \mathbf{x}_1^s(t; t_0, t') + O(\epsilon^2). \tag{2.4.5b}$$

Next, we calculate $\Delta_\epsilon^u(t'; t_0, t')$ and $\Delta_\epsilon^s(t'; t_0, t')$, and therefore $\Delta_\epsilon(t'; t_0, t')$, by obtaining the differentials $d\Delta_\epsilon^u(t; t_0, t')$ and $d\Delta_\epsilon^s(t; t_0, t')$ and integrating them from $t = -\infty$ to $t = t'$ and from $t = t'$ to $t = \infty$, respectively. To obtain expressions for the derivatives $\dot{\mathbf{x}}_1^u(t; t_0, t', \epsilon)$, $\dot{\mathbf{x}}_1^s(t; t_0, t', \epsilon)$ that appear in the expressions of these differentials, we differentiate Eqs. 2.4.3 with respect to time. Then we equate the resulting expressions of the derivatives $\dot{\mathbf{x}}^u(t; t_0, t', \epsilon)$, $\dot{\mathbf{x}}^s(t; t_0, t', \epsilon)$ with their expressions obtained, to first order, from Eq. 2.3.1 by expanding $\mathbf{f}(\mathbf{x}^u)$ and $\mathbf{f}(\mathbf{x}^s)$ in power series in the differences $(\mathbf{x}^u - \mathbf{x}_h)$ and $(\mathbf{x}^s - \mathbf{x}_h)$. The details of the calculations are given in Appendix A1. They yield the following expression for the distance function:

$$\Delta_\epsilon(t'; t_0, t') = \epsilon M(t_0) + O(\epsilon^2), \tag{2.4.6}$$

where $M(t_0)$ is given by Eq. 2.4.1.

Extensive mathematical studies of the Melnikov function (and, for systems with more degrees of freedom, its counterpart the Melnikov vector) are reported in Sanders (1982), Meyer and Sell (1989), and Chow and Yamashita (1992).

2.5 MELNIKOV FUNCTIONS FOR SPECIAL TYPES OF PERTURBATION. MELNIKOV SCALE FACTOR

In this section we specialize the expression for the Melnikov function for a number of types of perturbation of interest in applications. In Section 2.5.1 we consider systems with perturbation vector components $\epsilon g_m(\mathbf{x}, t) = \epsilon[\gamma_m(\mathbf{x}, t) + q_m(\mathbf{x})]$ $(m = 1, 2)$, that is, systems for which the excitation $\epsilon\gamma_m(\mathbf{x}, t)$ is state dependent.

In Section 2.5.2 we consider the cases of multiplicative excitation, that is, $\gamma_m(\mathbf{x}, t) = \gamma_m(\mathbf{x})P(t)$ $(m = 1, 2)$, and additive excitation, where $\gamma_m(\mathbf{x}) = \gamma_m = \text{const}$ $(m = 1, 2)$. We show that for these cases the fluctuating part of the Melnikov function can be viewed as the output of a linear filter with time-dependent input, and we define the Melnikov impulse response function.

In Section 2.5.3 we consider the case $\gamma_m(\mathbf{x}, t) = \gamma_m(\mathbf{x}) \,\mathrm{Re}[\exp(j(\omega t + \theta_0))]$ $(m = 1, 2)$, and the case of additive harmonic excitation, where $\gamma_m(\mathbf{x}) = \gamma_m = \text{const}\,(m = 1, 2)$. For these cases we define the Melnikov scale factor, which is shown in a subsequent section to be a measure of the dependence on frequency of an excitation's effectiveness in inducing transitions.

In Section 2.5.4 we consider the case $\gamma_m(\mathbf{x})\Sigma_i a_i \mathrm{Re}[\exp(j(\omega_i t + \theta_{0i}))]$ $(m = 1, 2)$, where, in general, the frequencies ω_i are incommensurate, that is, the case of multiplicative quasiperiodic excitation, and its additive counterpart, wherein $\gamma_m(\mathbf{x}) = \gamma_m = \text{const}\,(m = 1, 2)$. The quasiperiodic excitation case is used in Chapter 5 as a building block for developing the stochastic Melnikov method.

2.5.1 Melnikov Function for Systems with Perturbation Vector Components $\gamma_m(\mathbf{x}, t) + q_m(\mathbf{x})$ $(m = 1, 2)$

In many practical applications the perturbation vector of the system 2.3.1 has the form

$$\mathbf{g}(\mathbf{x}, t) = \{\gamma_1(\mathbf{x}, t) + q_1(\mathbf{x}), \gamma_2(\mathbf{x}, t) + q_2(\mathbf{x})\}. \qquad (2.5.1)$$

In Eq. 2.5.1 the excitation is *state dependent*, that is, it depends upon the system state $\mathbf{x}$.

From Eqs. 2.4.1 and 2.5.1 we obtain the corresponding expression for the Melnikov function:

$$\begin{aligned} M(t_0) &= \int_{-\infty}^{\infty} [f_1(\mathbf{x}_h(\zeta))q_2(\mathbf{x}_h(\zeta)) - f_2(\mathbf{x}_h(\zeta))q_1(\mathbf{x}_h(\zeta))]\, d\zeta \\ &\quad + \int_{-\infty}^{\infty} [f_1(\mathbf{x}_h(\zeta))\gamma_2(\mathbf{x}_h(\zeta), \zeta + t_0) \\ &\quad - f_2(\mathbf{x}_h(\zeta))\gamma_1(\mathbf{x}_h(\zeta), \zeta + t_0)]\, d\zeta \qquad (2.5.2a) \\ &= -k + \int_{-\infty}^{\infty} [f_1(\mathbf{x}_h(\zeta))\gamma_2(\mathbf{x}_h(\zeta), \zeta + t_0) \\ &\quad - f_2(\mathbf{x}_h(\zeta))\gamma_1(\mathbf{x}_h(\zeta), \zeta + t_0))]\, d\zeta, \qquad (2.5.2b) \end{aligned}$$

$$k = \int_{-\infty}^{\infty} [f_2(\mathbf{x}_h(\zeta))q_1(\mathbf{x}_h(\zeta)) - f_1(\mathbf{x}_h(\zeta))q_2(\mathbf{x}_h(\zeta))]\, d\zeta. \qquad (2.5.3)$$

In Section 2.5.2 we specialize Eq. 2.5.2b for systems with multiplicative excitation $\gamma_m(\mathbf{x}, t) = \gamma_m(\mathbf{x})P(t)$ $(m = 1, 2)$. In Sections 2.5.3 and 2.5.4 we further specialize Eq. 2.5.2b for the cases, needed for the development of the stochastic Melnikov approach, where $P(t)$ is harmonic and quasiperiodic.

2.5.2 Melnikov Function for Systems with Multiplicative and Additive Excitation. The Melnikov Function as Output of a Linear Filter

We now consider the case where, in Eq. 2.5.1,

$$\gamma_m(\mathbf{x}, t) = \gamma_m(\mathbf{x})P(t), \quad m = 1, 2. \tag{2.5.4}$$

In Eq. 2.5.4 the factors $\gamma_m(\mathbf{x})$ $(m = 1, 2)$ are functions of the state $\mathbf{x}$ of the system. The excitation 2.5.4 is referred to as *multiplicative*. In the particular case $\gamma_m(\mathbf{x}) = \gamma_m = \text{const}$ $(m = 1, 2)$ the excitation is referred to as *additive*. Equation 2.5.2b yields

$$\begin{aligned} M(t_0) &= -k + \int_{-\infty}^{\infty} [f_1(\mathbf{x}_h(\zeta))\gamma_2(\mathbf{x}_h(\zeta)) \\ &\quad - f_2(\mathbf{x}_h(\zeta))\gamma_1(\mathbf{x}_h(\zeta))]P(\zeta + t_0)\, d\zeta \qquad (2.5.5\text{a}) \\ &= -k + \int_{-\infty}^{\infty} [f_1(\mathbf{x}_h(-\zeta))\gamma_2(\mathbf{x}_h(-\zeta)) \\ &\quad - f_2(\mathbf{x}_h(-\zeta))\gamma_1(\mathbf{x}_h(-\zeta))]P(-\zeta + t_0)\, d\zeta \qquad (2.5.5\text{b}) \\ &= -k + \int_{-\infty}^{\infty} h(\zeta)P(t_0 - \zeta)\, d\zeta, \qquad (2.5.5\text{c}) \\ h(\zeta) &= f_1(\mathbf{x}_h(-\zeta))\gamma_2(\mathbf{x}_h(-\zeta)) - f_2(\mathbf{x}_h(-\zeta))\gamma_1(\mathbf{x}_h(-\zeta)). \qquad (2.5.5\text{d}) \end{aligned}$$

The integral of Eq. 2.5.5c is a *convolution integral*, and the time-dependent part of $M(t_0)$ is the result of the convolution of $P(t)$ with $h(t)$. Equation 2.5.5c is also written in the form

$$M(t_0) = -k + h * P, \tag{2.5.5e}$$

where the symbol $*$ denotes convolution.

Let us consider the particular case $P(t) \equiv \delta(t)$, where $\delta(t)$ is the Dirac delta function defined as

$$\delta(\tau) = 0 \quad \text{for} \quad \tau \neq 0; \qquad \lim_{\Delta t \to 0} \int_{-\Delta t/2}^{\Delta t/2} \delta(\tau)\, d\tau = 1.$$

Equation 2.5.5c yields $M(t_0) = -k + h(t_0)$. As shown by Eq. 2.5.5c, the time-dependent part of the Melnikov function is a linear superposition of elemental contributions $h(\zeta)P(t_0 - \zeta)\, d\zeta$, and may therefore be viewed as the *output* of a *Melnikov linear filter with input* $P(t_0)$. The function $h(\zeta)$ is called the *Melnikov impulse response function.* We note that the Melnikov linear filter is *noncausal.*[6] For material on linear filters see also Section 4.3.

Remark. Equation 2.5.5c shows that the expression for the Melnikov function has the same form for both multiplicative and additive excitation. The factors γ_m $(m = 1, 2)$ in the impulse response function $h(\zeta)$ depend on $\mathbf{x}$ if the excitation is multiplicative, and are constant if the excitation is additive (Eq. 2.5.5d). Note, however, that we do not consider the more general definitions of state-dependent excitation where $\gamma_m(\mathbf{x})$ also depends explicitly on the time t.

2.5.3 Melnikov Function for Systems with Excitation $\gamma_m(\mathbf{x}, t) = \gamma_m(\mathbf{x})\,\mathrm{Re}[\exp(j(\omega t + \theta_0))]$ $(m = 1, 2)$. Melnikov Scale Factor. Systems with Additive Harmonic Excitation. Examples

In this section we consider systems 2.3.1 with perturbation 2.5.1, and assume a state-dependent excitation 2.5.4 in which

$$P(t) = \mathrm{Re}[\exp j(\omega t + \theta_0)], \tag{2.5.6}$$

$(j = \sqrt{-1})$.[7] We show that the Melnikov function has the expression

$$M(t_0) = -k + |\boldsymbol{\alpha}(\omega)| \cos[\omega t_0 + \theta_0 - \psi(\omega)], \tag{2.5.7}$$

in which k is defined by Eq. 2.5.3,

$$\boldsymbol{\alpha}(\omega) = \int_{\infty}^{-\infty} h(\zeta) \exp(-j\omega\zeta)\, d\zeta, \tag{2.5.8}$$

$$C(\omega) = \int_{-\infty}^{\infty} h(\zeta) \cos(\omega\zeta)\, d\zeta, \tag{2.5.9a}$$

$$S(\omega) = \int_{-\infty}^{\infty} h(\zeta) \sin(\omega\zeta)\, d\zeta, \tag{2.5.9b}$$

$$|\boldsymbol{\alpha}(\omega)| = [C(\omega)^2 + S(\omega)^2]^{1/2}, \tag{2.5.10a}$$

$$\psi(\omega) = \tan^{-1}[S(\omega)/C(\omega)], \tag{2.5.10b}$$

[6] A causal filter is one whose output at time t vanishes for $\xi > t$. For example, if $P(\xi)$ is an impulsive load that induces a deflection $x(t)$ in a linearly elastic beam, it is clear that $x(t) \equiv 0$ for $t < \xi$, that is, the load $P(\xi)$ can cause no deflection *before* it is applied. In expressions involving outputs of a nonphysical nature (e.g., the Melnikov function), filters need not be causal.

[7] Throughout the text we denote complex quantities by bold italic letters. Unless otherwise indicated we denote the real part of a complex quantity z by $\mathrm{Re}(z)$ or z.

and $h(\zeta)$ is defined by Eq. 2.5.5d. The functions $\boldsymbol{\alpha}(\omega)$, $|\boldsymbol{\alpha}(\omega)|$, and $\psi(\omega)$ are referred to as the Melnikov transfer function, Melnikov scale factor, and Melnikov phase angle, respectively. In keeping with well-established usage, the real and imaginary parts of the function $\alpha(\omega)$ are denoted by $C(\omega)$ and $S(\omega)$, respectively (Eqs. 2.5.9).

Equation 2.5.7 is obtained as follows. Using Eq. 2.5.5c,

$$\boldsymbol{M}(t_0) = -k + \int_{-\infty}^{\infty} h(\zeta)\exp[j(\omega(-\zeta+t_0)+\theta_0)]\, d\zeta \tag{2.5.11}$$

$$= -k + \left\{\int_{-\infty}^{\infty} h(\zeta)\exp(-j\omega\zeta)\, d\zeta\right\}\exp[j(\omega t_0+\theta_0)], \tag{2.5.12}$$

where $\boldsymbol{M}(t_0)$ is a complex function whose real part is $M(t_0)$. We define $\boldsymbol{\alpha}(\omega)$ by Eq. 2.5.8, and obtain from Eqs. 2.5.8, 2.5.9, and 2.5.12

$$M(t_0) = -k + [C(\omega)\cos(\omega t_0 + \theta_0) + S(\omega)\sin(\omega t_0 + \theta_0)]. \tag{2.5.13}$$

Equations 2.5.10 imply $C(\omega) = |\boldsymbol{\alpha}(\omega)|\cos[\psi(\omega)]$, $S(\omega) = |\boldsymbol{\alpha}(\omega)|\sin[\psi(\omega)]$. Equation 2.5.7 follows from the substitution of these expressions in Eq. 2.5.13 and the application in the resulting expression of the identity $\cos a \cos b + \sin a \sin b = \cos(a - b)$.

Melnikov functions for systems with additive harmonic excitation correspond to the particular case $\gamma_m(\mathbf{x}) \equiv \gamma_m = \text{const}\ (m = 1, 2)$.

Example 2.5.1 *Melnikov function for the harmonically excited Duffing–Holmes oscillator*

We consider the Duffing–Holmes oscillator, that is, the system 2.3.1 in which

$$\mathbf{f}(\mathbf{x}) = \{x_2, -V'(x_1)\}, \tag{2.5.14a}$$

$$\mathbf{g}(\mathbf{x}, t) = \{0, \gamma\cos(\omega t) + \beta x_2\}, \tag{2.5.14b}$$

$$V(x_1) = -(a/2)x_1^2 + (b/4)x_1^4. \tag{2.5.14c}$$

As indicated by Eq. 2.5.14b, the perturbation includes a viscous damping term $q_2(x_1, x_2) = \beta x_2$. (For the particular case $a = b = 1$ the system is referred to as the standard Duffing–Holmes oscillator—see Eq. 2.1.7.) The equations of motion are

$$\dot{x}_1 = x_2, \tag{2.5.15a}$$

$$\dot{x}_2 = ax_1 - bx_1^3 + \epsilon[-\beta x_2 + \gamma\cos(\omega t)], \qquad a > 0, b > 0. \tag{2.5.15b}$$

The same procedure used to obtain Eq. 2.2.1 yields the homoclinic orbits

$$\{x_{h1}^{\pm}(t), x_{h2}^{\pm}(t)\} = \{\pm(2a/b)^{1/2}\,\mathrm{sech}(a^{1/2}t), \mp a(2/b)^{1/2} \times \mathrm{sech}(a^{1/2}t)\tanh(a^{1/2}t)\}. \quad (2.5.16)$$

Since $\gamma_1 = 0$, $\gamma_2 = \gamma = \text{const}$, and $f_1(\mathbf{x}_h(\zeta)) = x_{h2}(\zeta)$ is odd in ζ, it follows from Eqs 2.5.5d and 2.5.9a that $C(\omega)$ vanishes. The method of residues yields

$$S(\omega) = \gamma(2/b)^{1/2}\pi\omega\,\mathrm{sech}(\pi\omega/(2a^{1/2})). \quad (2.5.17)$$

Since $C(\omega) \equiv 0$, it follows from Eq. 2.5.10a that the Melnikov scale factor is $|\boldsymbol{\alpha}(\omega)| \equiv S(\omega)$. Using the expression for $\mathbf{x}_{h2}(\tau)$ (Eq. 2.5.16) and recalling that $d\tanh\tau = \mathrm{sech}^2\,\tau\,d\tau$, we obtain

$$k = -\beta\int_{-\infty}^{\infty} x_{h2}^2(\zeta)d\zeta = 4\beta a^{3/2}/(3b), \quad (2.5.18)$$

so the Melnikov function is

$$M(t_0) = -4\beta a^{3/2}/(3b) + \gamma(2/b)^{1/2}\pi\omega\,\mathrm{sech}(\pi\omega/(2a^{1/2}))\sin\omega t_0. \quad (2.5.19)$$

For the standard Duffing–Holmes oscillator ($a = b = 1$), $S(\omega)$ is plotted in Fig. 2.6 for $\gamma = 1$.

Example 2.5.2 *Melnikov function for the harmonically excited rf-driven Josephson junction*

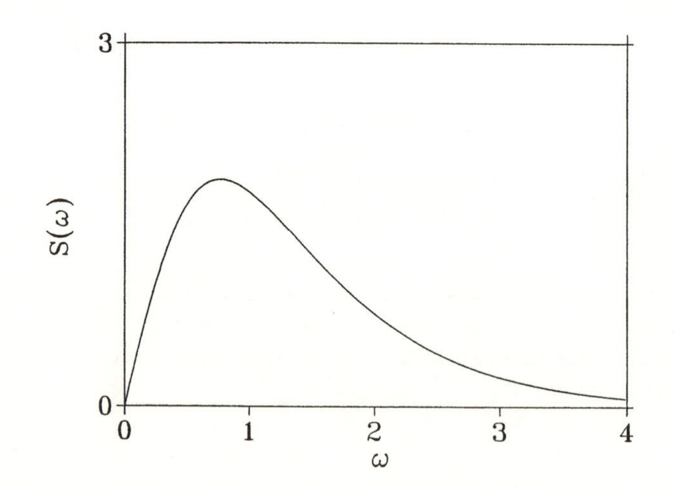

Figure 2.6. Melnikov scale factor for the standard Duffing–Holmes oscillator.

The equation for the rf-driven Josephson junction is

$$\dot{x}_1 = x_2 \tag{2.5.20a}$$

$$\dot{x}_2 = a \sin x_1 + \epsilon[b_0 + b_1 \sin \omega\tau - \beta x_2], \tag{2.5.20b}$$

where the constants b_0 and b_1 are the normalized amplitudes of the dc and radio frequency (rf) components of the driving current, and x_2 is the normalized voltage across the junction. Using the expression of the heteroclinic orbits of the system's Hamiltonian counterpart (Eq. 2.2.4), the Melnikov function can be shown to be

$$M(t_0) = 8\beta a^{1/2} - 2\pi\{b_0 + (b_1 \sin \omega t_0)/\cosh[\pi\omega/(2a^{1/2})]\} \tag{2.5.21}$$

(Genchev, Ivanov, and Todorov, 1983).

Remark. For systems with additive excitation and $\gamma_1 = 0, \gamma_2 = \gamma$ (e.g., Eqs. 2.5.15 or Eqs. 2.5.20), the Melnikov scale factor is usually defined as the function $|\boldsymbol{\alpha}(\omega)|$ (or, if $C(\omega) \equiv 0$, the function $S(\omega)$) corresponding to $\gamma = 1$. *Unless otherwise noted, we will henceforth use this definition for all systems with additive excitation and* $\gamma_1 = 0, \gamma_2 = \gamma$. Note that if this definition is adopted the factor γ must multiply the function $|\boldsymbol{\alpha}(\omega)|$ in the expression for the Melnikov function 2.5.7.

2.5.4 Melnikov Function for Systems with Excitation $\boldsymbol{\gamma_k(\mathbf{x}) \sum_i a_i \operatorname{Re}[\exp(j(\omega_i t + \theta_{0i}))]}$ $(k = 1, 2)$. Systems with Additive Quasiperiodic Excitation

In this section we obtain the expression of the Melnikov function for systems 2.3.1 with perturbation 2.5.1, and multiplicative excitation 2.5.4 in which

$$P(t) = \sum_{i=1}^{l} a_i \operatorname{Re}[\exp(j(\omega_i t + \theta_{0i}))], \tag{2.5.22}$$

where a_i $(i = 1, 2, \ldots, l)$ are constants, the frequencies ω_i are, in general, incommensurate, and $0 < \theta_{0i} \leq 2\pi$.

Steps similar to those that led to Eqs. 2.5.7 and 2.5.9 yield the result

$$M(t_0) = \sum_{i=1}^{l} a_i\{C(\omega_i)\cos(\omega_i t_0 + \theta_{0i}) + S(\omega_i)\sin(\omega_i t_0 + \theta_{0i})\} - k \tag{2.5.23a}$$

$$= \sum_{i=1}^{l} a_i |\boldsymbol{\alpha}(\omega_i)| \cos[\omega_i t_0 + \theta_{0i} - \psi(\omega_i)] - k. \tag{2.5.23b}$$

In Eq. 2.5.23, k is defined by Eq. 2.5.3, and $h(\zeta)$, $C(\omega_i)$, $S(\omega_i)$, $|\mathbf{a}(\omega_i)|$, and $\psi(\omega_i)$ are defined by Eqs. 2.5.5d, 2.5.9a, 2.5.9b, 2.5.10a, and 2.5.10b, respectively. Note that the constants a_i of Eqs. 2.5.22 and 2.5.23 are distinct from the functions γ_1, γ_2, that are incorporated into the Melnikov scale factor $|\boldsymbol{\alpha}(\omega_i)|$ (Eqs. 2.5.10a, 2.5.8, and 2.5.5d).

Melnikov functions for systems with additive quasiperiodic excitation correspond to the particular case $\gamma_m(\mathbf{x}) \equiv \gamma_m = \text{const}\,(m = 1, 2)$. If $\gamma_1 = 0$, $\gamma_2 = \gamma = \text{const}$, the Melnikov scale factor will be defined as the function $|\boldsymbol{\alpha}(\omega_i)|$ corresponding to $\gamma = 1$ (see Remark at end of Section 2.5.3).

2.6 CONDITION FOR THE INTERSECTION OF STABLE AND UNSTABLE MANIFOLDS. INTERPRETATION FROM A SYSTEM ENERGY VIEWPOINT

In this section we show that *if the Melnikov function has simple zeros* (has no zeros), *then the stable and unstable manifolds intersect transversely* (do not intersect). For a particular class of systems we interpret this condition from an energy viewpoint.

The following proposition is fundamental to Melnikov theory:

> If the Melnikov function has a simple zero for some t_0, for sufficiently small ϵ the stable and unstable manifolds of the perturbed systems, $W^s(\Gamma_\epsilon(t))$ and $W^u(\Gamma_\epsilon(t))$, intersect transversely.[8] If $M(t_0) \neq 0$ for all t_0, then
>
> $$W^s(\Gamma_\epsilon(t)) \cap W^u(\Gamma_\epsilon(t)) = \varnothing,$$
>
> that is, the manifolds do not intersect.

The first part of this proposition follows from Eq. 2.4.6, which shows that, for sufficiently small ϵ, if $M(t_0)$ has a simple zero so does the distance function. A simple zero of the distance function implies a change of the distance function's sign which, in turn, implies a reversal of the orientation of the vector $\mathbf{x}^u - \mathbf{x}^s$ (Eq. 2.4.2), and therefore a transverse intersection of the stable and unstable manifolds (Fig. 2.7). It also follows from Eq. 2.4.6 that, for sufficiently small ϵ, if $M(t_0)$ is bounded away from zero for all t_0, there is no intersection between the perturbed system's stable and unstable manifolds.

[8]A simple zero implies transverse intersection, whereas a double zero implies tangency.

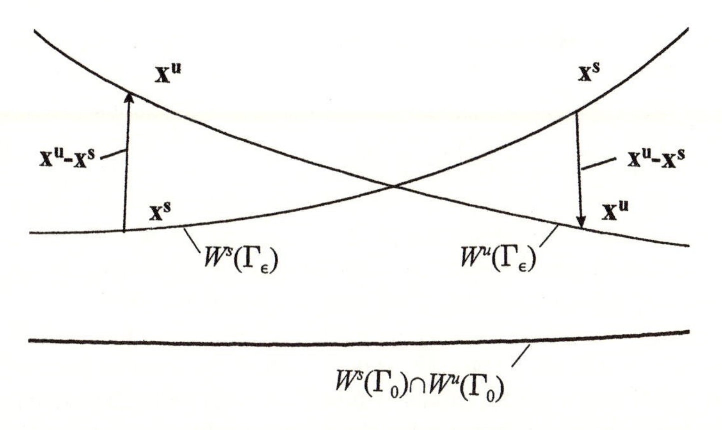

Figure 2.7. Plane of section through unperturbed manifolds $W^s(\Gamma_0) \cap W^u(\Gamma_0)$ and intersecting perturbed manifolds $W^s(\Gamma_\epsilon) \cap W^u(\Gamma_\epsilon)$, showing opposite orientations of the vector $x^u x^s$ on left and right sides of the manifolds' intersection point P'.

Let us now consider Eq. 2.3.1 in which $\mathbf{f}(\mathbf{x}) \equiv \{0, -V'(x_1)\}$, $\mathbf{g}(\mathbf{x}, t) \equiv \{0, -\beta x_2 + \gamma \cos[\omega(t + t_0)]\}$ $(\beta > 0)$, that is, the one-degree-of-freedom viscously damped harmonically forced system

$$\dot{x}_1 = x_2, \tag{2.6.1a}$$

$$\dot{x}_2 = V'(x_1) + \epsilon\{-\beta x_2 + \gamma \cos[\omega(t + t_0)]\}. \tag{2.6.1b}$$

Suppose the motion takes place on the unperturbed system's homoclinic orbit. If the motion occurs over a small distance δx_{h1} (as in Section 2.4.1 the subscript h designates coordinates of the homoclinic orbit), then the energy δE_{diss} lost during this motion is equal to the damping force times δx_{h1}, that is,

$$\delta E_{diss} = -\epsilon \beta x_{h2} \delta x_{h1}, \tag{2.6.2a}$$

and the total energy lost during motion over the entire homoclinic orbit is

$$E_{diss} = -\epsilon\beta \int_{-\infty}^{\infty} \dot{x}_{h1}\, dx_{h1} = -\epsilon\beta \int_{-\infty}^{\infty} \dot{x}_{h1}^2 dt. \tag{2.6.2b}$$

The energy gained during a motion over the entire homoclinic loop is equal to the work performed by the excitation, that is,

$$E_{exc} = \epsilon\gamma \int_{-\infty}^{\infty} \cos[\omega(t + t_0)]\, dx_{h1} = \epsilon\gamma \int_{-\infty}^{\infty} \cos[\omega(t + t_0)] \dot{x}_{h1}\, dt. \tag{2.6.3}$$

The energy contributed by the potential force during a motion over the entire homoclinic loop is zero.

The total energy produced during the motion over an entire homoclinic loop is therefore

$$E_{tot}=E_{diss}+E_{exc}=\epsilon\left\{-\beta\int_{-\infty}^{\infty}\dot{x}_{h1}^{2}\,dt+\gamma\int_{-\infty}^{\infty}\cos[\omega(t+t_0)]\dot{x}_{h1}\,dt\right\}. \quad (2.6.4)$$

It follows from Eqs. 2.6.1 and 2.6.4 that $E_{tot}=\epsilon M(t_0)$, where $M(t_0)$ denotes the Melnikov function (Eq. 2.4.1), and is therefore approximately equal to the distance function—see Eq. 2.4.6 (Tan and Radmore, 1995).

The condition $\max(E_{tot})>0$ implies that the maximum of the second term between braces in Eq. 2.6.4 is larger than the first term, that is, that the Melnikov function has simple zeros or, equivalently, that the stable and unstable manifolds intersect transversely. For the system 2.6.1 transverse intersections thus imply that the energy of the system can drive the motion over the potential barrier and out of a potential well (Fig. 2.1a).

In Section 2.7 we use geometric rather than energy considerations to show that this is true not only for the system 2.6.1 but for the more general class of planar systems defined by (2.3.1).

2.7 POINCARÉ MAPS, PHASE SPACE SLICES, AND PHASE SPACE FLUX

In this section we describe the structure of intersecting stable and unstable manifolds and explain how that structure makes possible the occurrence of transitions. We first consider periodically excited systems, and define Poincaré sections and Poincaré maps, which allow a considerable simplification in the study of the dynamics. For quasiperiodically excited systems a Poincaré map cannot be constructed, and we discuss the counterpart of the Poincaré section, the phase space slice. We then define and illustrate the calculation of the phase space flux factor, a measure of the frequency of transitions that has useful practical applications (Chapters 6 and 7).

2.7.1 Periodically Perturbed Systems. Poincaré Maps

2.7.1.1 Stable and Unstable Manifolds in a Periodically Excited System: Intersections with Poincaré Planes of Section

Let the excitation in Eq. 2.3.1 be periodic in time with period ω. As in Section 2.1.1, we denote by $\mathbf{x}(\mathbf{x}_0, t_0; t)$ the coordinates at time t of the solution of (2.3.1) whose coordinates at the time t_0 are $\mathbf{x}_0$. Let $t_0=t'+2k\pi/\omega$, where t' is fixed and k is an integer, and $t=t'+2(k+1)\pi/\omega$. Since

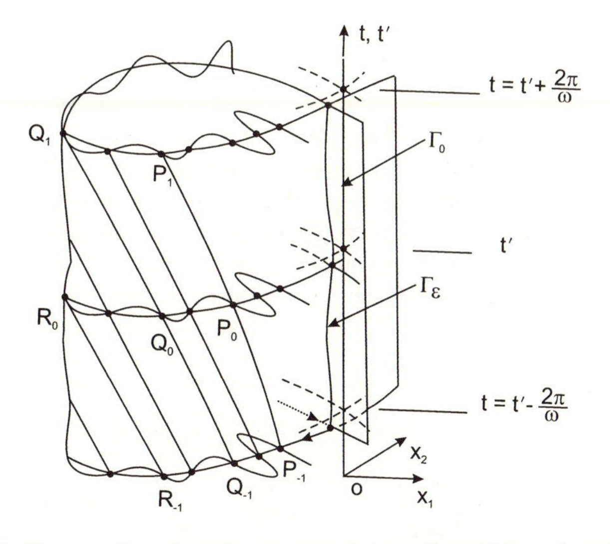

Figure 2.8. Plane sections through transverse intersecting stable and unstable manifolds for a system with excitation of period ω. Intersections of the manifolds with successive planes of section are identical. Intersections of unperturbed manifolds with the planes of section are partially shown near Γ_0 (dashed lines).

the periodic excitation remains unchanged if the integer k changes, it follows from Eq. 2.3.1 that, for fixed $\mathbf{x}_0$, the coordinates $\mathbf{x}(\mathbf{x}_0, t' + 2k\pi/\omega; t' + 2(k+1)\pi/\omega)$ $(k = 0, \pm 1, \pm 2, \ldots)$ do not depend on k.

In particular, we note that the coordinates $\mathbf{x}$ of the intersections of the curve Γ_ϵ (Fig. 2.8) with *Poincaré planes of section*—planes of section with elevations $t' + 2k\pi/\omega$, $k = 0, \pm 1, \pm 2, \ldots$—are identical for all k. This is also true of the intersections with those planes of (a) the local stable and unstable manifolds, and (b) the global stable and unstable manifolds. A depiction of identical intersections with planes of section $t' - 2\pi/\omega$, t', and $t' + 2\pi/\omega$ (t' fixed) is shown in Fig. 2.8.

Consider, in Fig. 2.8, the sequence of intersections of an orbit with successive planes $t' + 2k\pi/\omega$ $(k = 0, \pm 1, \pm 2, \ldots)$, $Q_{-1}, Q_0, Q_1, \ldots$ (forward motion), or $Q_1, Q_0, Q_{-1}, \ldots$ (reverse motion).[9] Since, for fixed $\mathbf{x}_0$, $\mathbf{x}(\mathbf{x}_0, t' + 2k\pi/\omega; t' + 2(k+1)\pi/\omega)$ are independent of k, point Q_1 may be identified with its orthogonal projection R_0 on the plane $t = t'$ (i.e., Q_1 and R_0 have the same coordinates $\mathbf{x}$), and point Q_{-1} may be identified with point P_0. Instead of the sequence Q_{-1}, Q_0, Q_1 on three successive planes of section, we may therefore consider the equivalent sequence P_0, Q_0, R_0 in a map represented on a single plane of section.

[9]Point Q_1 is referred to as the forward iterate (or, in the terminology of Poincaré (1892), the consequent) of Q_0. Point Q_{-1} is the backward iterate (the antecedent) of Q_0.

2.7.1.2 *Suspension of a Nonautonomous System in an Autonomous Phase Space. Definition of Poincaré Maps*

The reasoning applied in the preceding illustration to the set of orbits constituting the system's stable and unstable manifolds is applicable to any set of orbits of the system defined by Eq. 2.3.1 with time-periodic perturbation.

Let the perturbation period of that nonautonomous system be $T = 2\pi/\omega$. We now define that system's Poincaré map. To this end we suspend the system in the extended phase space (or autonomous phase space) $\{x_1, x_2, \theta\}$, where $\theta = [\omega t + \theta_0] \pmod{2\pi}$, that is, we write the equations of motion in the form

$$\begin{aligned} \dot{x}_1 &= f_1(x_1, x_2) + \epsilon g_1(x_1, x_2, \theta), \\ \dot{x}_2 &= f_2(x_1, x_2) + \epsilon g_2(x_1, x_2, \theta), \\ \dot{\theta} &= \omega. \end{aligned} \tag{2.7.1}$$

The flow generated by Eq. 2.7.1 is $\phi_t = \phi\{\mathbf{x}(t), \theta(t)\}$ (see Section 2.1.1). We denote by Σ the intersection of ϕ_t with the plane $\theta(t) = \theta_c = \text{const}$. The Poincaré map P: $\Sigma \to \Sigma$ is defined as

$$\phi\{\mathbf{x}[\theta_c - \theta_0]/\omega, \theta_c\} \to \phi\{\mathbf{x}[\theta_c - \theta_0 + 2\pi]/\omega, \theta_c + 2\pi\} \tag{2.7.2a}$$

or, on account of the periodicity of the perturbation,

$$\phi\{\mathbf{x}[\theta_c - \theta_0]/\omega, \theta_c\} \to \phi\{\mathbf{x}[\theta_c - \theta_0 + 2\pi]/\omega, \theta_c\}. \tag{2.7.2b}$$

Equations 2.7.2 are the mathematical expression for the flow ϕ_t of the observation made at the end of Section 2.7.1.1 for the particular case of the sequences represented in Fig. 2.8.

Remark 1. The coordinate θ is *bounded*. Therefore, if a set (x_1, x_2) is bounded in the phase plane, the set (x_1, x_2, θ) is bounded in the extended phase space. This observation is useful for the definition of the bounded sets referred to as attractors in the phase space $\{x_1, x_2, \theta\}$ (Section 3.1.3).

Remark 2. The fact that, in a system with periodic excitation, intersections of the stable and unstable manifolds with Poincaré planes of section are identical makes it possible to represent the cylindrical surfaces of Figs. 2.3 and 2.4 as the 2-torus T_0 and the 2-torus T_ϵ, respectively (Fig. 2.9). In the tori of Fig. 2.9 the counterparts of lines Γ_0 and Γ_ϵ (Figs. 2.3 and 2.4) are the 1-tori (closed curves) Γ_{T0} and $\Gamma_{T\epsilon}$, respectively.

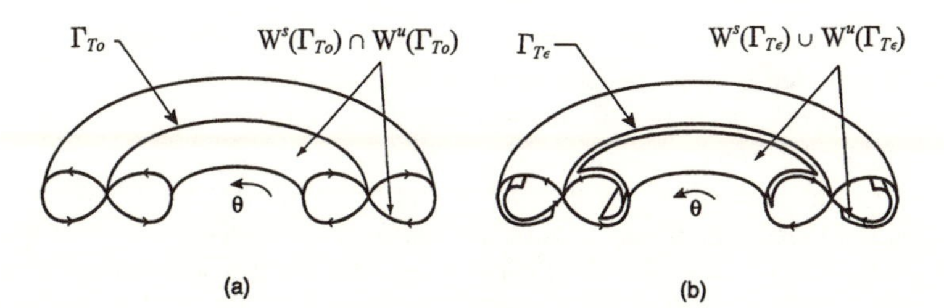

Figure 2.9. Half cut views of (a) 2-torus $T_0(\Gamma_{T0})$ representing manifolds of unperturbed system, and (b) 2-torus $T_\epsilon(\Gamma_{T\epsilon})$ representing separated manifolds of perturbed system.

2.7.2 Homoclinic Tangle and Transport across the Pseudoseparatrix

Note in Fig. 2.8 that since Q_0 is part of both the stable and unstable manifolds, so are Q_{-1} and its backward iterates, and Q_1 and its forward iterates. Therefore, if in a plane cross section of the stable and unstable manifolds, the manifolds intersect once, they intersect an infinity of times. The points of intersection are called *homoclinic points.*

Segments of manifolds intersecting transversely define *lobes*. Successive lobes in a Poincaré map represent sections by successive planes $t' + 2k\pi$ of the *same* three-dimensional lobe in the phase space (x_1, x_2, t). For example, the two-dimensional lobes containing the points B_{-2}, B_{-1}, B_0 in Fig. 2.10 are successive sections of the three-dimensional lobe at the right of points Q_{-1}, Q_0, Q_1 (Fig. 2.8).

The union of the perturbed manifold segments that most closely approximates the separatrix of the unperturbed system is called a *pseudoseparatrix.*

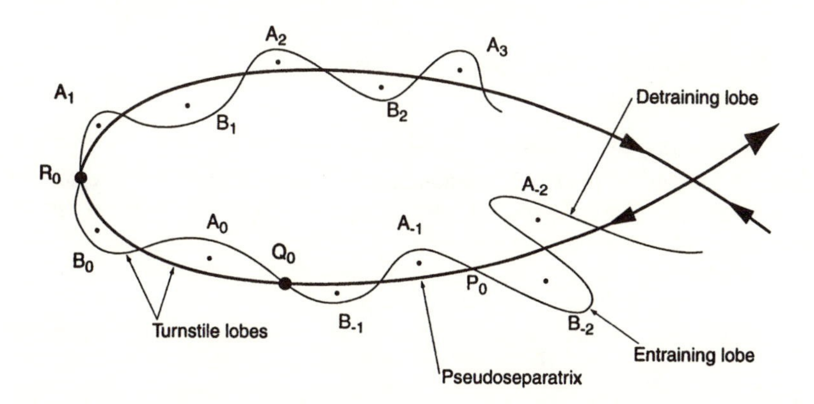

Figure 2.10. Poincaré map. Chaotic transport takes point A_{-2}, located within the core bounded by the pseudoseparatrix, through A_{-1} and A_0 to point A_1, located outside the core.

The intersecting manifolds form a *homoclinic tangle*. The interior of the region bounded by a pseudoseparatrix is referred to as the *core*.

Consider now, in Fig. 2.10, an orbit that starts at a point A_{-2} within a lobe and inside the pseudoseparatrix. The orbit must remain in the interior of a lobe. As was pointed out for the case of unperturbed systems in Section 2.2, this is implicit in the uniqueness of the system's solutions. If the orbit intersected the lobe—a manifold—at some point, that point would belong to two distinct solutions: one belonging to the manifold, and one crossing the manifold. Forward iteration carries A_{-2} to A_{-1} to A_0, which are still inside the core, then to A_1 and $A_2, \ldots$, which are *outside* the core. Similarly, a point B_{-2} contained by a lobe that is outside the core is iterated to B_{-1}, to B_0, and then to B_1 and $B_2, \ldots$, which are inside the core. The lobes that effect the transport from the interior to the exterior of the core, and from the exterior to the interior, are known as *entraining lobes* (e.g., those containing B_{-2}, B_{-1}) and *detraining lobes* (e.g., those containing A_{-2}, A_{-1}), respectively. The lobes containing A_0 and B_0 are referred to as *turnstile lobes* (MacKay, Meiss, and Percival, 1984; Wiggins, 1992).

We may now interpret the unperturbed system as one for which the Melnikov function is identically zero, rather than having simple zeros, as required by the necessary condition for the occurrence of transitions. We saw in Section 2.2 that the unperturbed system has a separatrix that is impermeable. On the other hand, the perturbed system for which the Melnikov function has simple zeros has a pseudoseparatrix that is permeable. The transport across the pseudoseparatrix is associated with periodically induced motions featuring transitions of the type illustrated in Fig. 1.2c. In Chapter 3 the chaotic character of the transport across the pseudoseparatrix and of the associated transitions is explained on theoretical grounds and illustrated numerically.

2.7.3 Autonomous Phase Space for Quasiperiodically Perturbed Systems

The quasiperiodically excited system defined by Eqs. 2.3.1, 2.5.1, 2.5.4, and 2.5.22 can be rendered autonomous by suspending it in the extended (autonomous) phase space $\{x_1, x_2, \theta_1, \theta_2, \ldots, \theta_i, \ldots, \theta_l\}$, where $\theta_i = [\omega_i t + \theta_{0i}](\text{mod}\, 2\pi)$, that is, by writing

$$\begin{aligned} \dot{\mathbf{x}} &= \mathbf{f}(\mathbf{x}) + \epsilon \mathbf{g}(\mathbf{x}, \theta), \\ \dot{\theta} &= \omega, \end{aligned} \tag{2.7.3}$$

$\theta = [\theta_1, \theta_2, \ldots, \theta_l]^{\mathrm{T}}$, $\omega = [\omega_1, \omega_2, \ldots, \omega_l]^{\mathrm{T}}$. We remark that, as was the case for the coordinate θ of the periodically excited system 2.7.1, the coordinates θ_i of the extended autonomous phase space are bounded.

2.7.4 Phase Space Slices

Let ω_l denote one of the frequencies ω_i of the quasiperiodic perturbation (Eq. 2.5.22). The coordinates $\mathbf{x}(\mathbf{x}_0, t' + 2(k+1)\pi/\omega_l; t' + 2k\pi/\omega_l)$ ($k = 0, \pm 1, \pm 2, \ldots$) (i.e., the coordinates at time $t' + 2(k+1)\pi/\omega_l$ of orbits starting from the same coordinate $\mathbf{x}_0$ at time $t' + 2k\pi/\omega_l$) are not independent of k, as was the case for the coordinates $\mathbf{x}(\mathbf{x}_0, t' + 2(k+1)\pi/\omega; t' + 2k\pi/\omega)$ in the periodically perturbed system. The system has no Poincaré map similar to Fig. 2.10 and, in particular, successive intersections of the stable and unstable manifolds with planes of section $t' + 2k\pi/\omega_l$ are not identical. Intersections of an orbit or set of orbits with planes $t' + 2k\pi/\omega_l$ are referred to as phase space slices.

Let us consider again for a moment the case of perturbation with period ω. We noted that if the time interval between planes of section is an integer multiple of the period $2\pi/\omega$, the intersections between those planes and the stable and unstable manifolds are identical. On the other hand, if the time interval between two planes of section differs from the perturbation period, the respective intersections with the manifolds, while no longer identical, are of the same topological type, as can be observed by inspection of Fig. 2.8.

This is also the case if the planes of section are phase space slices through the stable and unstable manifolds of a quasiperiodically perturbed system. Figure 2.11 (Beigie, Leonard, and Wiggins, 1991) shows an example of two consecutive phase space slices through the stable and unstable manifolds of a system with heteroclinic orbits perturbed by a quasiperiodic sum of two harmonics with frequencies ω_1 and ω_2. It was pointed out in Section 2.7.2 that successive two-dimensional lobes in a Poincaré map (Fig. 2.10) can be viewed as belonging to the same three-dimensional lobe in the phase space $\{x_1, x_2, t\}$. Similarly, in the consecutive phase space slices of Fig. 2.11, lobes with the same kind of shading belong to the same three-dimensional lobe in the phase space $\{x_1, x_2, t\}$.

We refer the reader to Appendix A2 for an instructive example of the numerical construction of phase space slices.

2.7.5 Phase Space Flux Factor

The phase space flux factor is defined as

$$\Phi = \lim_{\Gamma \to \infty} \frac{1}{2T} \int_{-T}^{T} [M(t_0)]^+ \, dt_0 \tag{2.7.4}$$

where $[f(t)]^+$ is the positive part of $f(t)$. The flux factor Φ increases as the excitation increases, and is a measure of a system's transport through the pseudoseparatrix. Applications of the flux factor to vessel capsizing and open-loop control are discussed in Chapters 6 and 7.

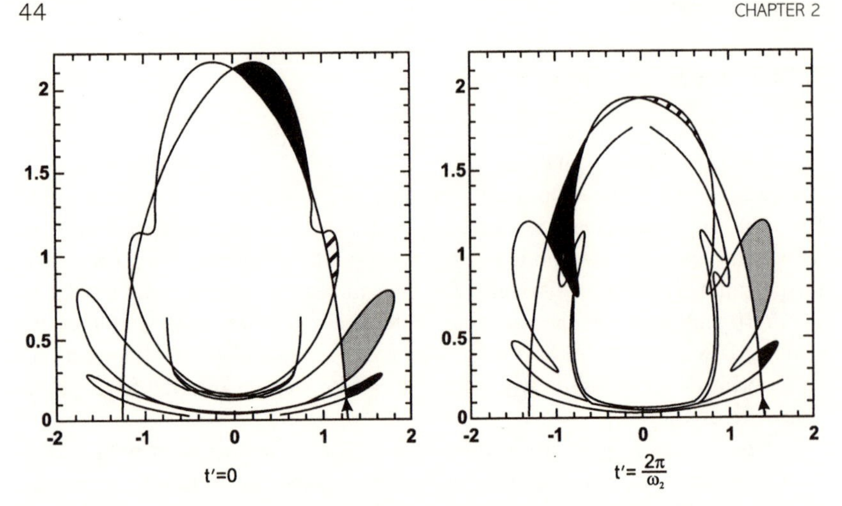

Figure 2.11. Two consecutive phase space slices through the stable and unstable manifolds of a quasiperiodically perturbed system whose unperturbed counterpart has heteroclinic orbits. Lobes with the same type of shading belong to the same three-dimensional lobes in the phase space $\{x_1, x_2, t\}$ (after Beigie, Leonard, and Wiggins, 1991).

Example 2.7.1 *Flux factor for harmonically excited system*

Consider a harmonically excited system 2.3.1 with vector field and perturbation given by Eqs. 2.5.14a,b, and assume that $x_{h2}(t)$ is an odd function of t, i.e., $C(\omega) = 0$. The expression for the system's Melnikov function is similar to the expression obtained in Example 2.5.1:

$$M(t_0) = a\cos(\omega\tau_0) - k \tag{2.7.5}$$

(Fig. 2.12a) where, using the notation of Example 2.5.1, $a = S(\omega)$ and $\tau_0 = t_0 - \pi/(2\omega)$. The phase space flux factor is

$$\Phi = \lim_{T\to\infty} \frac{1}{2T} \int_{-T}^{T} [a\cos(\omega\tau_0) - k]^+ \, d\tau_0. \tag{2.7.6}$$

Since the integrand is periodic it is sufficient to effect the integration and averaging over one period. The result is (Frey and Simiu, 1993)

$$\Phi = (1/\pi)S(\omega)\{1 - [k/S(\omega)]^2\} - (1/\pi)k\cos^{-1}[k/S(\omega)], \tag{2.7.7}$$

where the smallest positive value of the inverse cosine is used (Fig. 2.12b). Recall that we used the notation of Example 2.5 (Eq. 2.5.17), where $S(\omega)$ is proportional to the amplitude of the harmonic excitation γ. Figure 2.12b therefore shows that the phase space flux factor is a monotonically increasing function of γ.

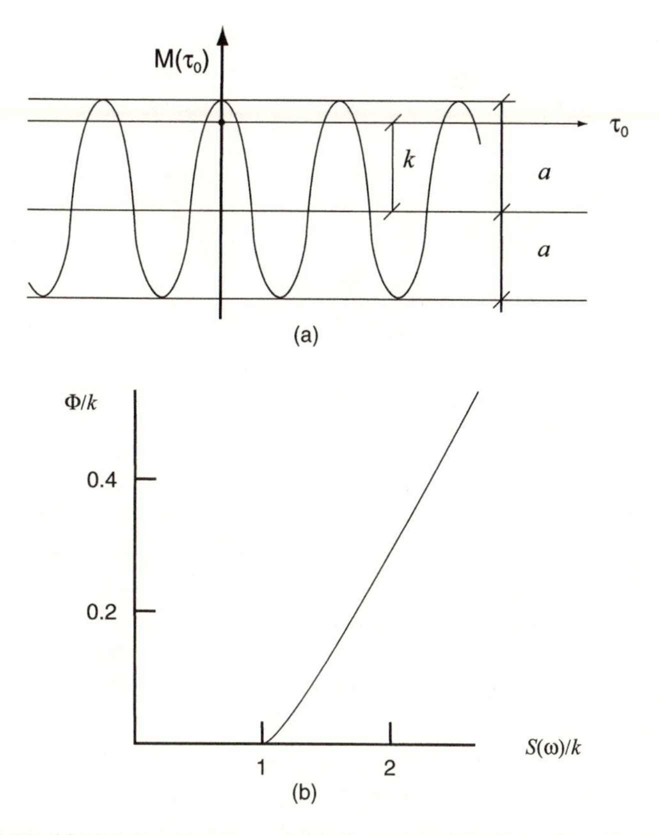

Figure 2.12. (a) Melnikov function and (b) phase space flux factor for a harmonically excited system.

2.8 SLOWLY VARYING SYSTEMS

In this section we consider the following class of slowly varying planar systems:

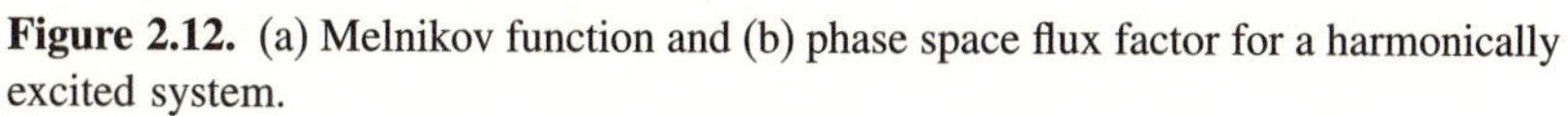

$$\begin{aligned} \dot{x} &= \frac{\partial}{\partial y} H(x, y, z) + \epsilon g_1(x, y, z, t; \boldsymbol{\mu}), \\ \dot{y} &= -\frac{\partial}{\partial x} H(x, y, z) + \epsilon g_2(x, y, z, t; \boldsymbol{\mu}), \\ \dot{z} &= \epsilon g_3(x, y, z, t; \boldsymbol{\mu}), \end{aligned} \tag{2.8.1}$$

where ϵ is small, the right-hand side is C^r differentiable ($r \geq 2$), $H(x, y, z)$ is a Hamiltonian with parameter z, and $\boldsymbol{\mu}$ is a vector of parameters. Following Wiggins and Holmes (1987, 1988), who considered the case of time-periodic perturbation functions, we present the derivation of the Melnikov function and the condition for the intersection of the stable and unstable manifolds. We then note that, for some slowly varying systems, transitions may occur by mechanisms other than transport associated with a homoclinic tangle (Whalen, 1996).

2.8.1 Unperturbed System

We first consider the unperturbed system obtained by setting $\epsilon = 0$ in Eq. 2.8.1:

$$\begin{aligned} \dot{x} &= \frac{\partial}{\partial y} H(x, y, z), \\ \dot{y} &= -\frac{\partial}{\partial x} H(x, y, z), \\ \dot{z} &= 0. \end{aligned} \tag{2.8.2}$$

Equation 2.8.2 represents a family of planar Hamiltonian systems with parameter z.

We assume that there exists an open interval J such that, for every $z \in J$, the associated planar system possesses a homoclinic orbit to a hyperbolic fixed point $p(z)$, with a homoclinic solution $\mathbf{x}_h^z(t)$ connecting p to itself, that is,

$$\lim_{t \to \pm\infty} \mathbf{x}_h^z(t) = p(z). \tag{2.8.3}$$

The condition for the unperturbed system's fixed point p to be hyperbolic is

$$\frac{\partial}{\partial x}\left(\frac{\partial H}{\partial y}\right)\frac{\partial}{\partial y}\left(\frac{\partial H}{\partial x}\right) + \frac{\partial}{\partial y}\left(\frac{\partial H}{\partial y}\right)\frac{\partial}{\partial x}\left(\frac{-\partial H}{\partial x}\right) > 0 \tag{2.8.4}$$

(see Section 2.1.1 between Eqs. 2.1.4a,b and 2.1.4c,d).

The fixed points of Eq. 2.8.2 are the solutions of the system

$$\frac{\partial}{\partial y} H(x, y, z) = 0, \tag{2.8.5a}$$

$$-\frac{\partial}{\partial x} H(x, y, z) = 0, \tag{2.8.5b}$$

which define x and y as functions of z. If (2.8.4) holds, the fixed points form in the $\{x, y, z\}$ phase space a curve $\Gamma(z)$ that depends smoothly on z and is the union of hyperbolic fixed points of the planar system 2.8.2 (Wiggins and Holmes, 1987). The normally hyperbolic one-dimensional invariant manifold $\Gamma(z)$ has two-dimensional stable and unstable manifolds denoted $W^s(\Gamma), W^u(\Gamma)$, respectively, such that their intersection

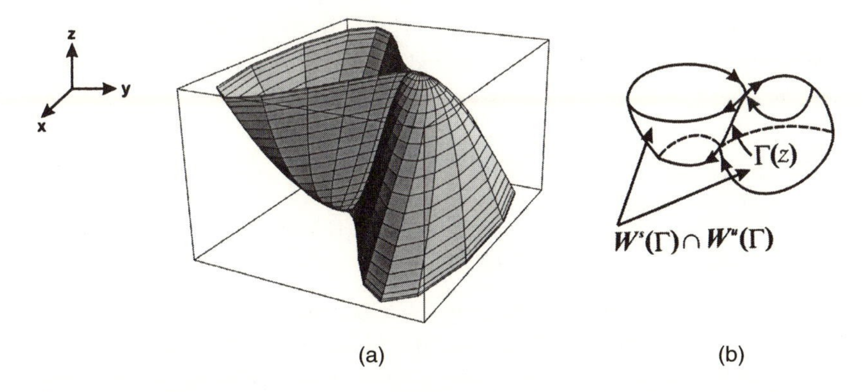

Figure 2.13. Unperturbed system's one-dimensional manifold $\Gamma(z)$ and stable and unstable manifolds $W^s(\Gamma)$ and $W^u(\Gamma)$: (a) full view; (b) partial view and notation (after Whalen, 1996).

$W^s(\Gamma) \cap W^u(\Gamma)$ is the union of the system's homoclinic orbits (Fig. 2.13). As was done in Fig. 2.3 within the three-dimensional phase space $\{x, y, t\}$, we construct within the *four-dimensional* phase space $\{x, y, z, t\}$ a cylinder whose directrix is the set $W^s(\Gamma) \cap W^u(\Gamma)$. Figure 2.13 represents an intersection of that cylinder with the plane $t =$ const.

2.8.2 Perturbed System. Melnikov Function

As was the case for planar systems, under assumptions similar to those listed for Eq. 2.3.1, including the assumption that the perturbations are sufficiently small, it can be shown that the two-dimensional surface $\{\Gamma(z), t\}$ persists in the perturbed system along with its stable and unstable manifolds, that is, the perturbed system has a normally hyperbolic invariant manifold

$$\Gamma_\epsilon(z, t; \epsilon) = \{\Gamma(z) + O(\epsilon), t\} \tag{2.8.6}$$

for all $z \in J$. $\Gamma_\epsilon(z, t; \epsilon)$ is as smooth as $\{\Gamma(z), t\}$ and has local stable and unstable manifolds, denoted $\{W^s_{loc}(\Gamma_\epsilon), t\}$ and $\{W^u_{loc}(\Gamma_\epsilon), t\}$, respectively, which are C^r-close to the local stable and unstable manifolds of the unperturbed system $\{W^s_{loc}(\Gamma), t\}$ and $\{W^u_{loc}(\Gamma), t\}$. For any fixed t, unlike $\{\Gamma(z), t\}$ which consisted only of fixed points, $\Gamma_\epsilon(z, t; \epsilon)$ consists of points with motions

$$\dot{z} = \epsilon g_3(\Gamma(z)) + O(\epsilon^2). \tag{2.8.7}$$

Equation 2.8.7 follows from the third of Eqs. 2.8.1 and Eq. 2.8.6 (for simplicity, in Eq. 2.8.7 and subsequent equations, we omit the parameters $\boldsymbol{\mu}$, and we denote $\{\Gamma(z), t\}$ by $\Gamma(z)$).

Suppose $g_3(\Gamma(z)) = \sum_{j=1}^{N} g_{3j}(x, y, z, t)$ $(j = 1, 2, \ldots, N)$, where each of the functions g_{3j} is periodic with period T_j. We now consider the averaged system

$$\dot{z} = \epsilon \overline{g_3(\Gamma(z))}, \tag{2.8.8}$$

$$\overline{g_3(\Gamma(z))} = \sum_{j=1}^{N} \frac{1}{T_j} \int_0^T g_{3j}(\Gamma(z), t)\, dt. \tag{2.8.9}$$

It follows from the averaging theorem (see Guckenheimer and Holmes, 1986, p. 167 for $N = 1$, and Verhulst, 1990, p. 154 for $N > 1$) that, if there exists $z_0 \in J$ such that the system averaged with respect to time has a hyperbolic fixed point

$$\overline{g_3(\Gamma(z_0))} = 0, \qquad d\overline{g_3(\Gamma(z_0))}/dz \neq 0, \tag{2.8.10a,b}$$

then the perturbed system has a hyperbolic orbit $\Gamma_\epsilon(z_0, t; \epsilon) = \{\Gamma(z_0) + O(\epsilon), t\}$. Trajectories with initial conditions on the surface $\Gamma_\epsilon(z, t; \epsilon)$ are attracted or repelled by the orbit $\Gamma_\epsilon(z_0, t; \epsilon)$ for, respectively,

$$d\overline{g_3(\Gamma(z_0))}/dz < 0 \quad \text{and} \quad d\overline{g_3(\Gamma(z_0))}/dz > 0. \tag{2.8.11a,b}$$

For the case 2.8.11a, a cross section by the plane $t = t_0$ through the four-dimensional phase space $\{x, y, z, t\}$ of the perturbed system yields the three-dimensional phase space slice shown in Fig. 2.14. (In the particular case of periodic excitation a Poincaré map may be constructed in the three-dimensional space $\{x, y, z\}$.) In the three-dimensional phase space slice, if (2.8.11a) holds, the intersections with a plane of section of the perturbed stable and unstable manifolds $W^s(\Gamma_\epsilon(z_0, t_0; \epsilon))$ and $W^u(\Gamma_\epsilon(z_0, t_0; \epsilon))$ are two- and one-dimensional, respectively, and motions with initial conditions on $\Gamma_\epsilon(z, t_0; \epsilon)$ are attracted by $\Gamma_\epsilon(z_0, t_0; \epsilon)$ (this case is represented in Fig. 2.14); if (2.8.11b) holds the stable and unstable manifolds are one- and two-dimensional, respectively, and motions with initial conditions on $\Gamma_\epsilon(z, t_0; \epsilon)$ are repelled by $\Gamma_\epsilon(z_0, t_0; \epsilon)$.

Also shown in Fig. 2.14 is the plane π passing through a point q_0 with coordinate z_0 of the unperturbed homoclinic orbit. The plane π is parallel to the z axis and is normal at point q_0 to the unperturbed homoclinic orbit passing though that point, that is, it contains the unit vector $\mathbf{e}_z$ parallel to the z axis, and the vector $\{\partial H(q_0)/\partial x, \partial H(q_0)/\partial y, 0\}$, which is normal to the vector $\{-\partial H(q_0)/\partial y, \partial H(q_0)/\partial x, 0\}$. (By virtue of Eqs. 2.8.2 the latter vector is tangent at q_0 to the unperturbed homoclinic orbits.) In addition, Fig. 2.14 shows the point q_ϵ^u representing the intersection with the plane π of the unstable manifold $W^u(\Gamma_\epsilon(z_0, t_0; \epsilon))$, and the point q_ϵ^s representing the

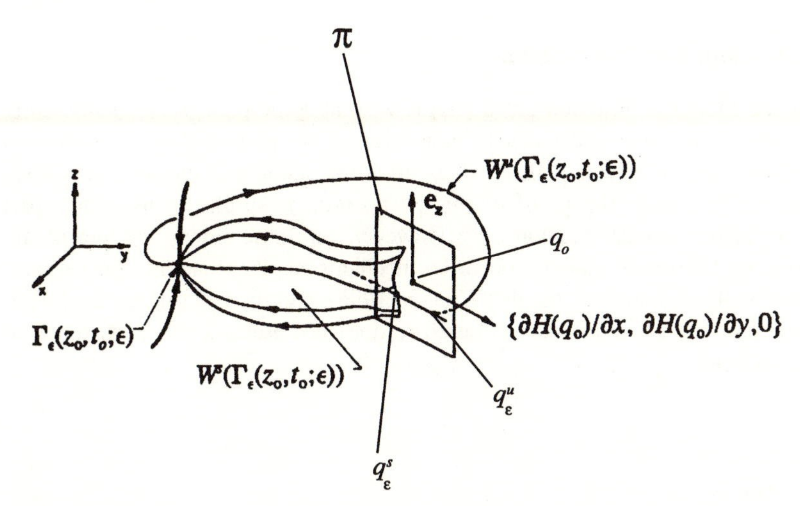

Figure 2.14. Space phase slice through perturbed manifolds (after Wiggins and Shaw, 1988).

intersection with the plane π of the orbit contained in the stable manifold $W^s(\Gamma_\epsilon(z_0, t_0; \epsilon))$ which has the same coordinate z as the point q^u_ϵ.

The Melnikov distance is the projection on the vector $\{\partial H(q_0)/\partial x, \partial H(q_0)/\partial y, 0\}$ of the vector $q^u_\epsilon - q^s_\epsilon$ and can therefore be written as

$$d = \frac{\{\partial H(q_0)/\partial x, \partial H(q_0)/\partial y, 0\} \cdot \{q^u_\epsilon - q^s_\epsilon\}}{[(\partial H(q_0)/\partial x)^2 + (\partial H(q_0)/\partial y)^2]^{1/2}}. \tag{2.8.12}$$

We have $q^u_\epsilon - q^s_\epsilon = \epsilon(\partial q^u_\epsilon/\partial\epsilon|_{\epsilon=0} - \partial q^s_\epsilon/\partial\epsilon|_{\epsilon=0}) + O(\epsilon^2)$. We define the Melnikov function as

$$M(q_0) = \{\partial H(q_0)/\partial x, \partial H(q_0)/\partial y, 0\} \cdot \{\partial q^u_\epsilon/\partial\epsilon|_{\epsilon=0} - \partial q^s_\epsilon/\partial\epsilon|_{\epsilon=0}\}. \tag{2.8.13}$$

Following calculations similar to those of Section 2.4.1, Eq. 2.8.13 yields

$$M(t_0) = \int_{-\infty}^{\infty} \nabla H(q_0^{z_0}(t_0)) \cdot \mathbf{g}(q_0^{z_0}(t, t + t_0)\, dt - \partial H(\Gamma(z_0))/\partial z \int_{-\infty}^{\infty} g_3(q_0^{z_0}(t, t + t_0)\, dt, \tag{2.8.14}$$

where $\nabla H = \{\partial H(q_0)/\partial x, \partial H(q_0)/\partial y, \partial H(q_0)/\partial z\}$, $\mathbf{g} = \{g_1, g_2, g_3\}$, and $q_0^{z_0}(t)$ is the unperturbed homoclinic orbit connecting the hyperbolic fixed point of the unperturbed system at elevation z_0 to itself (Wiggins and Holmes, 1987, 1988).

2.8.3 Manifold Intersections

If the Melnikov function has a simple zero at some point t_0, for sufficiently small ϵ the stable and unstable manifolds of the perturbed system intersect transversely. If $M(t_0) \neq 0$ for all t_0, then the stable and unstable manifolds do not intersect. The proof of this proposition is similar to its counterpart for planar systems (Section 2.6). However, for some slowly varying planar systems transitions can also occur by mechanisms other than transport associated with a homoclinic tangle (Whalen, 1996), and the fact that the Melnikov function has no simple zeros therefore does not preclude the occurrence of transitions.

Chapter Three

Chaos in Deterministic Systems and the Melnikov Function

In Chapter 2 we considered deterministic planar systems capable of having a Melnikov function, and showed that the necessary condition for the occurrence of motions with transitions—motions of the type illustrated in Fig. 1.2c—is that their Melnikov function have simple zeros. In this chapter we establish that motions with transitions that occur if the Melnikov function has simple zeros are chaotic. Chaos implies, among others, three properties of the motion. First, the motion is sensitive to initial conditions. This entails motion unpredictability, since initial conditions can be ascertained with only limited precision. To the unavoidable errors in their specification there correspond differences between predicted and observed results that can be large even if those errors are small. Second, the motion appears to be irregular. Third, associated with the chaotic motion are geometrical structures with noninteger—fractal—dimensions.

In Section 3.1 we define and illustrate elementary mathematical objects needed in this chapter: Lyapounov exponents, which provide a measure of sensitivity to initial conditions, attractors, and basins of attraction. In Section 3.2 we introduce and illustrate Cantor sets, and discuss fractal dimensions.

In Section 3.3 we describe and examine in detail a landmark in dynamical systems theory: the two-dimensional Smale horseshoe map and its attendant Cantor set. An ingenious indexation system allows the construction, in the space of infinite sequences whose terms are either 0 or 1, of a map topologically equivalent to the Smale horseshoe map: the shift map.

In Section 3.4 we introduce symbolic dynamics techniques and use them to analyze the shift map. The analysis makes it possible to define chaos mathematically and establish that the dynamics of the shift map is chaotic. The equivalence between the shift map and the Smale horseshoe map implies that the dynamics on the Cantor set associated with the latter is chaotic as well.

In Section 3.5 we state the Smale-Birkhoff theorem. The theorem uses the results obtained on the dynamics under the Smale horseshoe map to establish that the necessary condition for the occurrence of chaotic motion in a planar system is that its Melnikov function have simple zeros. Such chaotic motion is precisely the motion with transitions made possible by intersecting stable and unstable manifolds. It is referred to as *homoclinic chaos*. The transitions and transport associated with it are referred to, respectively, as *chaotic transitions* and *chaotic transport.* If the Melnikov function has no simple zeros the motion cannot be chaotic—and cannot exhibit transitions. We also discuss *transient chaos*, that is, chaos in transient rather than steady-state motion, and its relation to the Melnikov function. We then present examples of motion corresponding to Melnikov functions with no zeros and with simple zeros.

In Section 3.6 we mention briefly an extension of the Smale horseshoe map that accommodates slowly varying planar systems. This extension is used to establish that the necessary condition for chaos is that the Melnikov function have simple zeros.

In Section 3.7 we describe a mechanical device capable of producing chaotic motion, and typical experimental results.

3.1 SENSITIVITY TO INITIAL CONDITIONS AND LYAPOUNOV EXPONENTS. ATTRACTORS AND BASINS OF ATTRACTION

3.1.1 Illustration of Sensitivity to Initial Conditions in One-Dimensional Maps. Stretching and Folding

In Section 3.3 it is shown that, in the two-dimensional Smale horseshoe map, sensitivity to initial conditions is associated with three geometrical features, stretching (expansion), contraction, and folding. As a preliminary to the Smale horseshoe map, we examine the dynamics of two one-dimensional maps, the tent map and a particular case of the logistic map. In particular, we show that for the tent map sensitivity to initial conditions is related in a transparent manner to two of those geometrical features: stretching and folding. For the logistic map we present a simple and direct mathematical proof of sensitivity to initial conditions.

Example 3.1.1 *The tent map*

The tent map is defined by the difference equation

$$x_{n+1} = 1 - 2|x_n - 1/2| \tag{3.1.1}$$

applied on the unit interval $0 \leq x_n \leq 1$. Equation 3.1.1 may be rewritten as

$$x_{n+1} = 2x_n, \quad 0 \leq x_n < 0.5; \qquad x_{n+1} = 2(1 - x_n), \quad 0.5 \leq x_n \leq 1.0, \tag{3.1.2a,b}$$

and is represented graphically in Fig. 3.1a,b. Note that all iterates x_{n+1}, $x_{n+2}, \ldots$ are contained in $[0, 1]$.

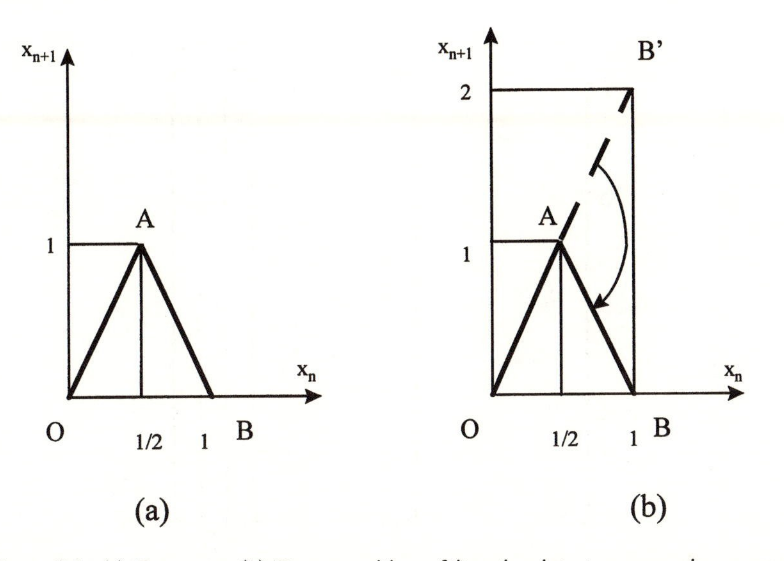

Figure 3.1. (a) Tent map. (b) Decomposition of iteration into two succssive operations: (1) expansion of segment OA into segment OB'; folding of segment AB' into segment AB.

The mapping may be viewed as entailing two stages. The first stage represents a stretching of the interval [0, 1] by a factor of 2. The second stage is a folding that keeps the iterate within the interval [0, 1] (Fig. 3.1). The stretching and folding processes just described are represented for three successive iterates in Figs. 3.2a, 3.2b, and 3.2c. The first iteration, from the state x_n to the state x_{n+1}, maps the interval $a_n b_n$ (i.e., the interval $[0, 0.25]$) on the interval $a_{n+1} b_{n+1}$ (i.e., $[0, 0.5]$); the interval $(0.25, 0.5]$ on $(0.5, 1]$; the interval $(0.5, 0.75]$ on $(1, 0.5]$; and the interval $(0.75, 1]$ on $(0.5, 0]$. After two iterations the state x_n is transformed as shown in Fig. 3.2c.

The stretching creates sensitivity to initial conditions, since the distance between two trajectories with given initial separation increases by a factor of 2 at each iteration, that is, the trajectories diverge exponentially. Although the divergence is limited by the folding to the size of the interval [0, 1], the stretching and folding can cause the distance between iterates of trajectories with small initial separation to vary erratically within that interval. For example, the reader may wish to use a pocket calculator to verify this statement for two trajectories with initial coordinates 0.258 and 0.259. After 10 iterations the respective coordinates are 0.192 and 0.784. Since, in practice, initial conditions can be known with only limited precision, the information needed to determine just where a trajectory will land after a sufficiently large number of iterations is not available, that is, the motion is unpredictable.

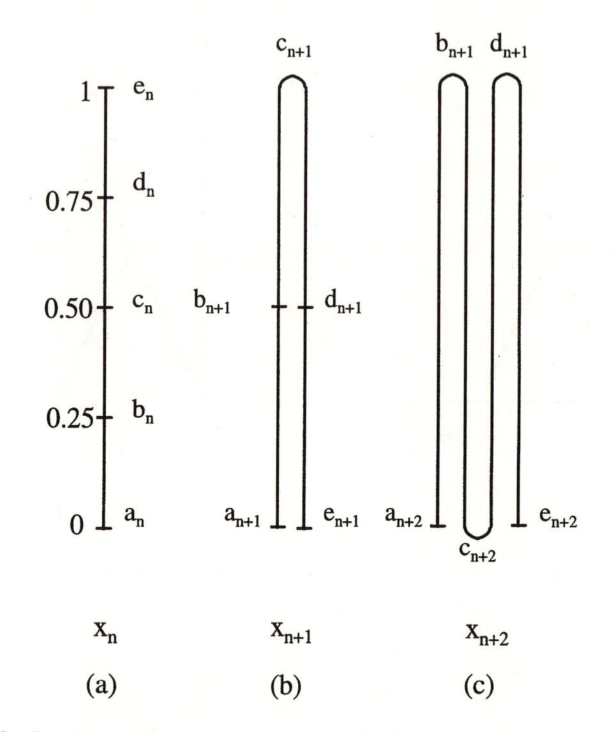

Figure 3.2. Combined effects of stretching and folding for three successive iterations of the tent map. The length of the interval $a_n - e_n$ is unity. Note the stylized horseshoe shape resulting from the stretching and folding.

Example 3.1.2 *The logistic map*

For a particular case of the logistic map it is possible to produce a simple mathematical proof of motion unpredictability. The equation of the logistic map is

$$x_{n+1} = 4cx_n(1 - x_n), \tag{3.1.3}$$

$0 \leq x_n \leq 1$. We consider the particular case $c = 1$. It follows from Eq. 3.1.3 that $0 \leq x_{n+1} \leq 1$. We now show that the map is sensitive to initial conditions. To do so we use the transformation $x_n = \sin^2 \pi\theta_n$. Equation 3.1.3 then becomes

$$\sin^2 \pi\theta_{n+1} = \sin^2 2\pi\theta_n. \tag{3.1.4}$$

Therefore

$$\theta_{n+1} = 2\theta_n, \tag{3.1.5}$$

from which it follows that

$$\theta_n = 2^n \theta_0. \tag{3.1.6}$$

Since we can add any integer to θ_0 without changing the value of x_0, we have

$$\theta_n = 2^n \theta_0 \pmod 1. \tag{3.1.7}$$

For example, consider the initial condition $\theta_0 = 0.8046875\cdots = 1/2 + 1/4 + 1/32 + 1/64 + 1/128 + \cdots$ and its successive iterates (the base 2 equivalents are given in parentheses)

$$\begin{aligned}
\theta_0 &= 1/2+1/4+0+0+1/32+1/64+1/128+\cdots & (\theta_0 &= 0.1100111\ldots),\\
\theta_1 &= 1/2+0+0+1/16+1/32+1/64+\cdots & (\theta_1 &= 0.100111\ldots),\\
\theta_2 &= 0+0+1/8+1/16+1/32+\cdots & (\theta_2 &= 0.00111\ldots),\\
\theta_3 &= 0+1/4+1/8+1/16+\cdots & (\theta_3 &= 0.0111\ldots),\\
\theta_4 &= 1/2+1/4+1/8+\cdots & (\theta_4 &= 0.111\ldots),\\
\theta_5 &= 1/2+1/4+\cdots & (\theta_5 &= 0.11\ldots),\\
\theta_6 &= 1/2+\cdots & (\theta_6 &= 0.1\ldots).
\end{aligned}$$

Note that, in base 2, an iteration amounts to shifting the "decimal" point to the right. This feature also appears in the *shift map*, which is discussed in Section 3.3 in conjunction with the Smale horseshoe map.

It is clear that since θ_0 can be known with only finite precision, no prediction whatever is possible on the position within [0, 1] of sufficiently high iterates of θ_0. For our example this is true of the seventh and higher iterates $\theta_7, \theta_8, \ldots$.

The reader can verify by using a pocket calculator that, for $c = 1$ and most initial conditions within the unit interval, iterates of Eq. 3.1.3 wander with no discernible pattern, in a seemingly erratic fashion. If θ_0 is changed by a small amount ϵ, θ_n changes by $\Delta\theta_n = 2^n\epsilon$ (Eq. 3.1.7), so the separation of initially close trajectories grows exponentially with time. However, as shown by Eq. 3.1.3 and as was the case for the tent map, the separation is limited to the size of the interval [0, 1].

3.1.2 Lyapounov Exponents

For a one-dimensional map, the Lyapounov exponent is defined as the average rate of exponential separation $\lambda = \lim_{n\to\infty,\epsilon\to 0}(1/n)\ln(\Delta\theta_n/\epsilon)$, where $\Delta\theta_n$ is the separation at iteration n between trajectories with initial separation

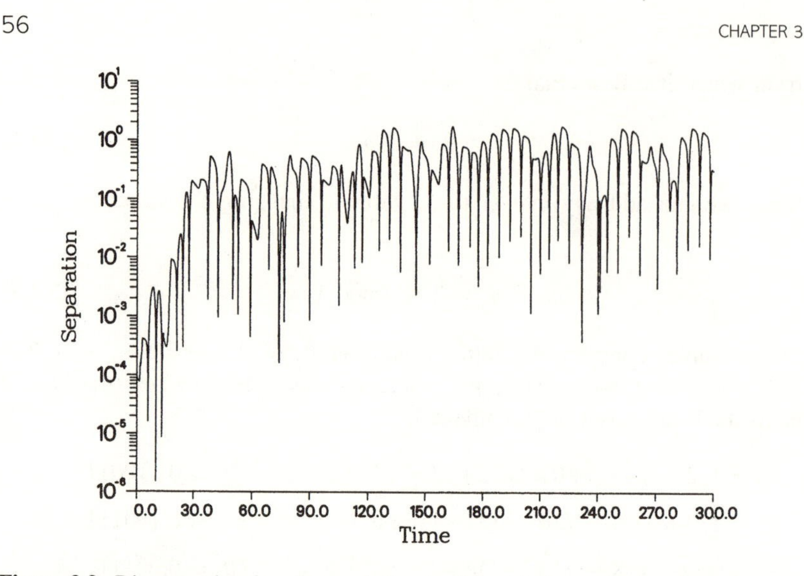

Figure 3.3. Diagram showing the evolution in time of the separation between two trajectories with 10^{-4} initial separation. On average the separation grows exponentially until an upper limit—the size of the attractor (see Section 3.1.3)—is reached (after Simiu and Cook, 1991).

ϵ. For the case of the logistic map with $c = 1$ it follows from this definition that $\Delta\theta_n/\epsilon = \exp(n\lambda)$, and from the result $\Delta\theta_n = 2^n\epsilon$ that $\lambda = \ln 2$. A positive Lyapounov exponent implies sensitivity to initial conditions.

An n-dimensional flow has n Lyapounov exponents, and sensitivity to initial conditions occurs if the largest Lyapounov exponent $\lambda_{\max}$ is positive (see, e.g., Bergé, Pomeau, and Vidal, 1984). In the particular case of continuous systems with known equations of motion, $\lambda_{\max}$ can be estimated by plotting the increase with time of an initially small separation, as in Fig. 3.3 (Simiu and Cook, 1991). Denoting the separation at time T by Δ, we have, by analogy with the discrete case,

$$\lambda_{\max} = \lim_{\substack{T\to\infty \\ \epsilon\to 0}} (1/T)\ln(\Delta/\epsilon). \tag{3.1.8}$$

For Fig. 3.3 the initial separation and the separation after time $T \approx 40$ are $\epsilon = 10^{-4}$ and $\Delta \approx 0.5$, respectively, so for this system $\lambda_{\max} \approx 0.2$. Note in Fig. 3.3 that after time $T = 40$ the separation Δ no longer increases. This is because after that time the steady-state motion is confined to a bounded region, just as was the case for the motion in the tent map.

For continuous systems whose equations are unknown, procedures have been developed that allow the estimation of Lyapounov exponents from information contained in observed time series. To circumvent computational problems associated with the exponential increase of the distance Δ with time, and with the limitations imposed by saturation illustrated in Fig. 3.3, the separation may be reset at regular, relatively small intervals, and an average obtained of the dilations over all the separate intervals (see, e.g., Benettin et al., 1976; Lichtenberg and Lieberman, 1992, Section 5.3). Eckmann and Ruelle (1992) warned that estimates of Lyapounov exponents have serious limitations insofar as the length requirements for the observed time series can be prohibitive.

3.1.3 Attractors and Basins of Attraction

Dissipative flows are characterized by the existence of attractors, that is, bounded sets approached asymptotically by trajectories originating from sets of initial conditions that constitute the attractors' respective *basins of attraction*. Consider, for example, a single-degree-of-freedom linear oscillator with low viscous damping and harmonic excitation. Its steady-state displacement x and velocity $\dot{x}$ are both harmonic. Therefore, if represented in the phase plane $\{x, \dot{x}\}$, the motion approaches a closed curve—a periodic attractor. (The orbits represented in the $\{x, \dot{x}, t\}$, phase space are not bounded, but, rather, spiral to infinity in the direction of the time coordinate. However, if the system is rendered autonomous, the system's attractor is bounded in the extended phase space, as indicated in Sections 2.7.1 and 2.7.3.) The basin of attraction in this example is the whole phase plane, that is, all trajectories, regardless of initial conditions, converge on the system's periodic attractor.

As a second example, consider an unforced system with small linear viscous damping evolving in the double-well potential of Fig. 2.1a. Depending upon the initial conditions, the motion approaches asymptotically the bottom of one or the other of the two wells. The system's two attractors are the fixed points C and C' in the system's phase plane; see Fig. 2.1b (the fixed points are attracting foci and correspond to complex eigenvalues, with negative real parts, of the characteristic equation associated with the linearized system). The system has two basins of attraction: the set of initial conditions of the trajectories that settle on the attractor C, and the set corresponding to the attractor C' (Fig. 3.4). Section 3.5 discusses in some detail examples of nonlinear systems with point attractors, periodic attractors (limit cycles), and chaotic attractors.

3.2 CANTOR SETS. FRACTAL DIMENSIONS

3.2.1 Cantor Sets

In the limit of an infinite number of iterations, a class of iterative processes result in infinite sets of disconnected points known as Cantor sets. We introduce these sets as a preliminary to the subsequent study of the Smale

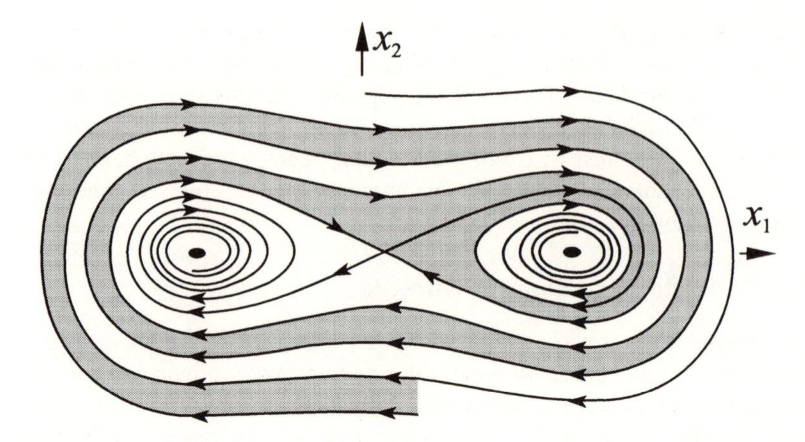

Figure 3.4. Point attractors and their basins of attraction for an unforced, dissipative double-well oscillator (after Thompson and Stewart, 1986).

horseshoe map. We limit ourselves to presenting, as a simple illustration, the middle-thirds Cantor set on the unit interval. The first step of the iterative process that defines it consists of removing the middle third of the interval. Each subsequent step consists of removing the middle thirds of the remaining intervals of the preceding step (Fig. 3.5). The middle-thirds Cantor set is the infinite set of disconnected points of the unit interval that subsist after the nth step as $n \to \infty$.

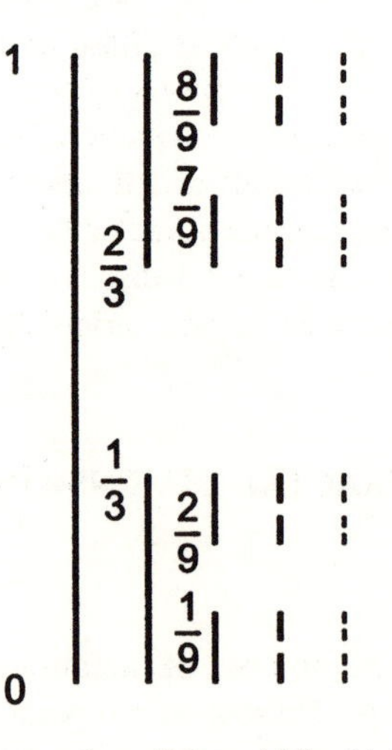

Figure 3.5. Construction of the middle-thirds Cantor set.

3.2.2 Fractal Dimensions

Cantor sets are among the geometrical objects referred to as fractals. Mandelbrot (1986) defined fractals as "shapes made of parts similar to the whole in some way." For the middle-thirds Cantor set the similarity entails missing middle thirds at all scales. Fractals have noninteger dimensions called fractal dimensions. Such dimensions are defined by expressions that, for the particular cases of nonfractal objects such as a point (a geometrical object with zero dimension), a line (a geometrical object with one dimension), a surface (a geometrical object with two dimensions), and volumes in a three-dimensional, $\ldots$, n-dimensional Euclidian space, yield the dimensions $0, 1, 2, 3, \ldots, n$, respectively. One such expression, known as the Hausdorff-Besicovitch dimension, is the limit, if that limit exists,

$$d = \lim_{\epsilon \to 0}\{\ln[N(\epsilon)]/\ln(1/\epsilon)\}, \tag{3.2.1}$$

where $N(\epsilon)$ is the minimum number of hypercubes of side ϵ needed to cover the object in question. By the definition 3.2.1 the dimension of the unit cube in the Cartesian space x, y, z is $\lim_{\epsilon\to 0}\{\ln[(1/\epsilon^3)]/\ln(1/\epsilon)\} = 3$.

For the middle-thirds Cantor set we obtain the fractal dimension d as follows. The initial set consisting of the unit interval can be covered by $N(\epsilon) = 1$ segment (one-dimensional cube) of size $\epsilon = 1$. For the first step (i.e., for the unit interval minus the middle third), $\epsilon = 1/3$ and $N(\epsilon) = 2$. For the second step $\epsilon = 1/9 = 1/3^2$ and $N(\epsilon) = 4 = 2^2$. For the nth step $\epsilon = 1/3^n$ and $N(\epsilon) = 2^n$. From Eq. 3.2.1, $d = \lim_{n\to\infty}\{(\ln 2^n)/(\ln 3^n)\} = 0.6309$. From the point of view of dimensionality the middle-thirds Cantor set may be viewed as intermediate between a point (dimension 0) and a line (dimension 1). The Cantor set, and in general objects with fractal dimension, are called fractal sets. For a thorough treatment of fractals see Feder (1988).

3.3 THE SMALE HORSESHOE MAP AND THE SHIFT MAP

In this section we describe the famous Smale horseshoe map and the shift map associated with it. In the limit of an infinite number of iterations the Smale horseshoe map results in a two-dimensional Cantor set denoted by Λ that is invariant under iteration, meaning that the iterate of any point belonging to Λ will also belong to Λ. We also show that the dynamics under the Smale horseshoe map on the set Λ is equivalent to the dynamics under the shift map on the set of bi-infinite sequences of the symbols 0 and 1, denoted by Σ_2. In Section 3.4 we show that the dynamics under the shift map of the set Σ_2 is chaotic. The equivalence of the shift map and the Smale horseshoe map therefore implies that the dynamics on Λ under the Smale horseshoe map is chaotic as well. This result is used in Section 3.5 to show that the necessary condition for chaos in a planar multistable system is that its Melnikov function have simple zeros.

We consider the two-dimensional Smale horseshoe map f defined by the following expressions

$$f: \begin{cases} \begin{pmatrix} x \\ y \end{pmatrix} \to \begin{pmatrix} \lambda & 0 \\ 0 & \mu \end{pmatrix} \begin{pmatrix} x \\ y \end{pmatrix}, & 0 \le x \le 1; 0 \le y \le 1/\mu, \quad (3.3.1a) \\ \begin{pmatrix} x \\ y \end{pmatrix} \to \begin{pmatrix} -\lambda & 0 \\ 0 & -\mu \end{pmatrix} \begin{pmatrix} x \\ y \end{pmatrix} + \begin{pmatrix} 1 \\ \mu \end{pmatrix}, & 0 \le x \le 1; \\ & 1 - 1/\mu \le y \le 1, \quad (3.3.1b) \end{cases}$$

where $\mu > 2$ and $\lambda < 1/2$. Equation 3.3.1a is an affine transformation that maps, for example, the rectangle H_0 to the rectangle V_0, that is,

$$f(H_0) = V_0 \tag{3.3.2}$$

(Fig. 3.6). Equation 3.3.1b entails, in addition to an affine transformation, a rotation by 180°, and a translation. (The two-dimensional matrix of Eq. 3.3.1b is the product of a two-dimensional rotation matrix and the two-dimensional affine transformation matrix of Eq. 3.3.1a.) For example, Eq. 3.3.1b takes the rectangle H_1 to V_1, that is,

$$f(H_1) = V_1 \tag{3.3.3}$$

(Fig. 3.6). The simultaneous iteration by Eqs. 3.3.1 that takes the horizontal rectangles H_0 and H_1 to V_0 and V_1, respectively, may be viewed as resulting from the following successive operations: (i) *contracting and stretching* of the unit square of Fig. 3.6a, yielding the strip of Fig. 3.6b, and (ii) *folding* of that strip into the stylized horseshoelike shape of Fig. 3.6c. The reader may verify Eqs. 3.3.2 and 3.3.3 by applying the mappings 3.3.1a and 3.3.1b to points a, b, c, d and e, f, g, h, respectively.

We now consider the *second iterate*, that is,

$$f^2(H_0 \cup H_1) \equiv f(f(H_0 \cup H_1)) = f(V_0 \cup V_1). \tag{3.3.4}$$

(The superscript denotes second iteration, rather than being an exponent.) The domain of f does not include the strip $1/\mu \le y \le 1 - 1/\mu$—see Eqs. 3.3.1 and Fig. 3.6a. For this reason we carry out the iteration $f(V_0 \cup V_1)$ in the following four steps. First, Eq. 3.3.1a is applied to the area $0 \le x \le \lambda$, $0 \le y \le 1/\mu$, that is, the area $H_0 \cap V_0$, yielding

$$f(H_0 \cap V_0) = V_{00} \tag{3.3.5}$$

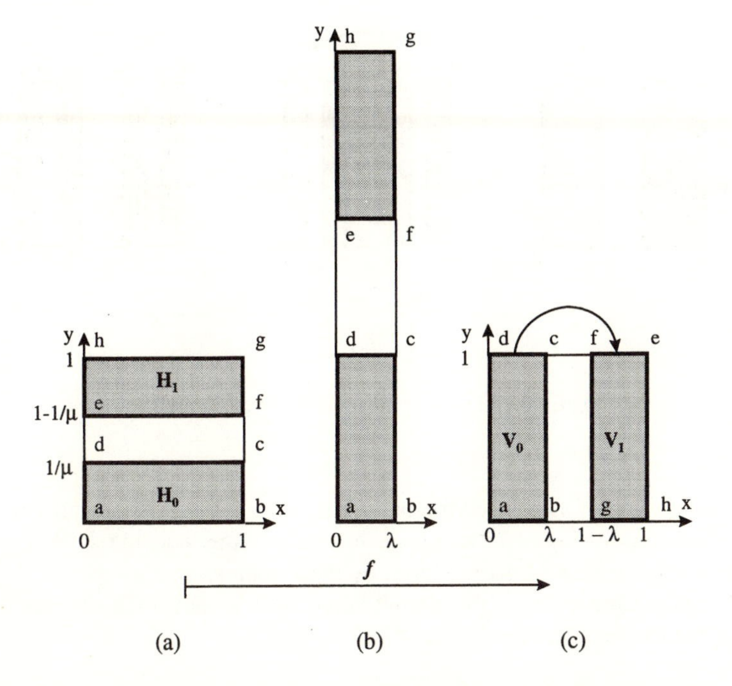

Figure 3.6. (a) Unit square containing rectangles H_0 and H_1; (b) contracting and stretching of unit square; (c) folding of structure depicted in (b), resulting in rectangles V_0 and V_1. Note positions of sides *ab* and *hg* at stages (a), (b), and (c).

(Figs. 3.6 and 3.7). Second, Eq. 3.3.1b is applied to the area $0 \leq x \leq \lambda$, $1 - 1/\mu \leq y \leq 1$, that is, the area $H_1 \cap V_0$. This yields

$$f(H_1 \cap V_0) = V_{10}. \tag{3.3.6}$$

Third, Eq. 3.3.1b yields

$$f(H_1 \cap V_1) = V_{11}. \tag{3.3.7}$$

Fourth, Eq. 3.3.1a yields

$$f(H_0 \cap V_1) = V_{01}. \tag{3.3.8}$$

The reader may check Eqs. 3.3.5 to 3.3.8 by applying Eqs. 3.3.1 to the corners of the rectangles being mapped and following their trajectories, as was suggested for points *a* through *d* and *e* through *h* for Fig. 3.6. Using Eqs. 3.3.5 to 3.3.8 we can write the second iterate as

$$f^2(H_0 \cup H_1) \equiv V_{00} \cup V_{10} \cup V_{11} \cup V_{01}. \tag{3.3.9}$$

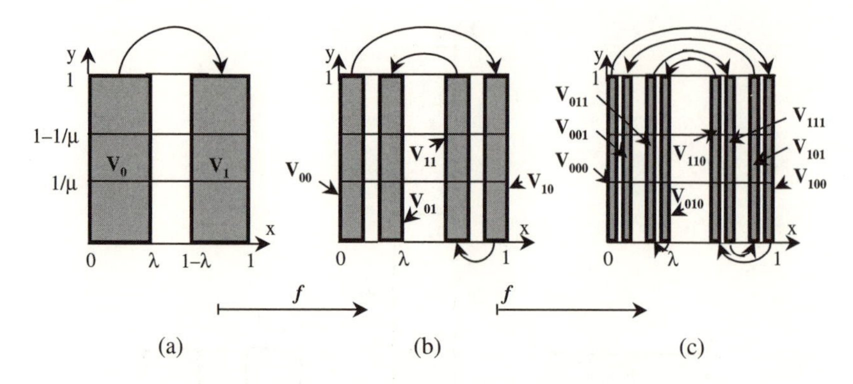

Figure 3.7. Forward iterations of areas V_0 and V_1.

The succession of steps used to obtain Eqs. 3.3.5 to 3.3.8 is indicated by the curved arrows of Fig. 3.7b and leads naturally to the indexing system being used for the vertical strips produced by the mapping.

The procedure used for the second iterate also applies for subsequent iterates. For the *third iterate*

$$f^3(H_0 \cup H_1) \equiv f(V_{00} \cup V_{10} \cup V_{11} \cup V_{01}) \qquad (3.3.10)$$

we have

$$\begin{aligned} f(H_0 \cap V_{00}) &= V_{000}, \quad f(H_1 \cap V_{00}) = V_{100}, \\ f(H_1 \cap V_{10}) &= V_{110}, \quad f(H_0 \cap V_{10}) = V_{010}, \end{aligned} \qquad (3.3.11\text{a,b,c,d})$$

$$\begin{aligned} f(H_0 \cap V_{11}) &= V_{011}, \quad f(H_1 \cap V_{11}) = V_{111}, \\ f(H_1 \cap V_{01}) &= V_{101}, \quad f(H_0 \cap V_{01}) = V_{001} \end{aligned} \qquad (3.3.11\text{e,f,g,h})$$

(Fig. 3.7). It is seen that the indexing system identifies each vertical strip obtained in the kth iteration of the area $H_0 \cup H_1$ by a sequence of k binary digits. We denote these sequences by $s_{-1}s_{-2}\cdots s_{-k}$, where $s_{-i} \in S$, $S = \{0, 1\}$ (i.e., the set S has two elements, the digits 0 and 1), and $i = 1, 2, \ldots, k$. The width of each vertical strip obtained in the kth iteration is λ^{k+1}. As $k \to \infty$, the width of the vertical strips approaches zero, that is, the strips become vertical lines, each of which is defined by an infinite sequence of binary digits $s_{-1}s_{-2}\cdots s_{-k}\cdots$.

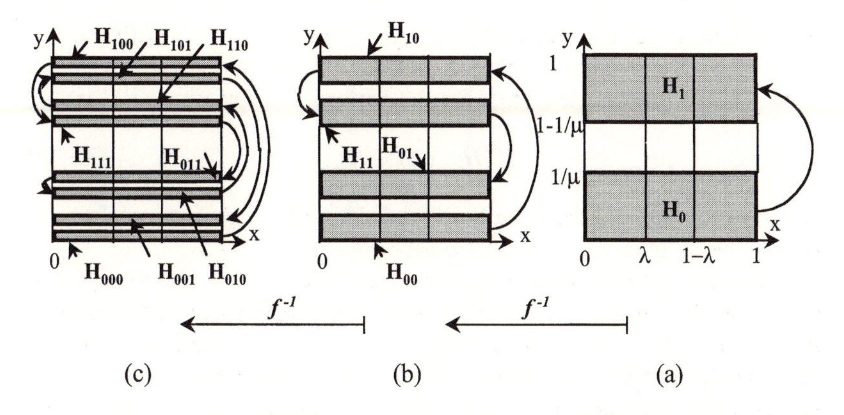

Figure 3.8. Reverse iterations of areas H_0 and H_1.

The *inverse* f^{-1} of the map f is immediately obtained from Eqs. 3.3.1:

$$f^{-1}: \begin{cases} \begin{pmatrix} x \\ y \end{pmatrix} \to \begin{pmatrix} 1/\lambda & 0 \\ 0 & 1/\mu \end{pmatrix} \begin{pmatrix} x \\ y \end{pmatrix}, & 0 \le x \le \lambda; \\ & 0 \le y \le 1, \quad (3.3.12a) \\ \begin{pmatrix} x \\ y \end{pmatrix} \to \begin{pmatrix} -1/\lambda & 0 \\ 0 & -1/\mu \end{pmatrix} \begin{pmatrix} x \\ y \end{pmatrix} + \begin{pmatrix} 1/\lambda \\ 1 \end{pmatrix}, & 1-\lambda \le x \le 1; \\ & 0 \le y \le 1. \quad (3.3.12b) \end{cases}$$

By Eqs. 3.3.12, the first and second iterations of the inverse map f^{-1} applied to the areas H_0 and H_1 yield the horizontal strips shown in Fig. 3.8b and 3.8c, respectively. The patterns for the indexing systems identifying horizontal strips and vertical strips are similar. For example,

$$f^{-1}(V_0 \cap H_0) = H_{00}, \quad f^{-1}(V_1 \cap H_0) = H_{10},$$
$$f^{-1}(V_1 \cap H_1) = H_{11}, \quad f^{-1}(V_0 \cap H_1) = H_{01}. \qquad (3.3.13)$$

From Eqs. 3.3.13 it follows that

$$f^{-1}(H_0 \cup H_1) = H_{00} \cup H_{10} \cup H_{11} \cup H_{01}. \qquad (3.3.14)$$

If the mapping f is applied to both sides of Eqs. 3.3.13, we obtain

$$f(H_{ij}) = V_i \cap H_j \in H_j, \quad i = 1, 2. \qquad (3.3.15)$$

The first iteration of the inverse map f^{-1} applied to $H_0 \cup H_1$ yielded the four horizontal strips of Eq. 3.3.13. We now consider the horizontal strips yielded by the kth iteration. Each of these horizontal strips is identified by a sequence of $k+1$ binary digits denoted by $s_k \cdots s_1 s_0$, where $s_i \in S$, $S = \{0, 1\}$, $i = 0, 1, \ldots, k$. The height of each strip is $(1/\mu)^{k+1}$. As can be verified in Fig. 3.8 for $k = 2$, the counterpart of Eq. 3.3.15 for a sequence $s_k \cdots s_1 s_0$ is

$$f^k(H_{s_k \cdots s_1 s_0}) \in H_{s_0}. \tag{3.3.16}$$

As $k \to \infty$, the strips become horizontal lines, and each horizontal line is defined by an infinite sequence of binary digits $\cdots s_k \cdots s_1 s_0$.

We are interested in the dynamics under the map f of the two-dimensional Cantor set Λ of points p representing intersections of vertical and horizontal lines identified by the infinite sequences $\cdots s_{-k} \cdots s_{-2} s_{-1}$ and $s_0 s_1 \cdots s_k \cdots$, respectively. Since to each point $p \in \Lambda$ there corresponds a bi-infinite sequence of 0's and 1's, $\cdots s_{-k} \cdots s_{-2} s_{-1} \cdot s_0 s_1 s_2 \cdots s_k \cdots$, where the "decimal" point separates past iterates from future iterates, there exists a one-to-one map

$$\phi(p) = \cdots s_{-k} \cdots s_{-2} s_{-1} \cdot s_0 s_1 s_2 \cdots s_k \cdots . \tag{3.3.17a}$$

that takes a point from the set Λ to a point in the space of sequences $\ldots s_{-k} \ldots s_{-2} s_{-1} \cdot s_0 s_1 s_2 \ldots s_k$.

It can be shown that ϕ is a homeomorphism, that is, in addition to being one to one, ϕ is onto[1] and continuous, and has a continuous inverse (Wiggins, 1988, p. 91). One iteration of $p \in \Lambda$ by f increases the number of past iterates by one, that is, to the point $f(p)$ of the set Λ there corresponds in the map ϕ the sequence $\cdots s_{-k} \cdots s_{-2} s_{-1} s_0 \cdot s_1 s_2 \cdots s_k \cdots$, so

$$\phi(f(p)) = \cdots s_{-k} \cdots s_{-2} s_{-1} s_0 \cdot s_1 s_2 \cdots s_k \cdots . \tag{3.3.17b}$$

We denote by Σ_2 the collection of all bi-infinite sequences on two symbols

$$s = \{\cdots s_{-k} \cdots s_{-2} s_{-1} \cdot s_0 s_1 s_2 \cdots s_k \cdots\}, s_i \in S \quad \text{for all } i \tag{3.3.18a}$$

$S = \{0, 1\}$. A sequence s may be viewed as a point in the sequence space Σ_2. The shift map σ is a map of Σ_2 into itself such that

$$\sigma(s) = \{\cdots s_{-k} \cdots s_{-2} s_{-1} s_0 \cdot s_1 s_2 \cdots s_k \cdots\}, \tag{3.3.18b}$$

i.e., σ acts on a sequence s by shifting its "decimal" place to the right. From the definition of $\sigma(s)$ and Eqs. 3.3.17 and 3.3.18a it is clear that

$$\phi(f(p)) = \sigma(\phi(p)). \tag{3.3.19a}$$

[1]A function $f(x)$ is *one to one* if $f(x) \neq f(y)$ whenever $x \neq y$. Let the domain and range of $f(x)$ be denoted by I and J. The function f is *onto* if for any $y \in J$ there is an x in I such that $f(x) = y$.

Using the notation $\phi \circ f \equiv \phi(f)$ (the symbol "$\circ$" denotes the composition of two functions), we rewrite Eq. 3.3.19a as

$$\phi \circ f(p) = \sigma \circ \phi(p). \tag{3.3.19b}$$

Since ϕ is a homeomorphism, by Eq. A3.1 (Appendix A3) the maps $f(p)$ and $\sigma(s)$ are topologically conjugate and therefore have equivalent dynamics. It follows that topological information on the dynamics of f can be obtained by studying the dynamics of σ.

Before concluding this section we show that all periodic points of the Smale horseshoe map are of the saddle type. This follows immediately from the application to Eq. 3.3.1 of the procedure described in Section 2.1.2, which yields the result that the eigenvalues of the matrix associated with the periodic points of period k are $|\lambda_{1k}| = \lambda^k < 1$ and $|\lambda_{2k}| = \mu^k > 1$.

3.4 SYMBOLIC DYNAMICS. PROPERTIES OF THE SPACE Σ_2. SENSITIVITY TO INITIAL CONDITIONS OF THE SMALE HORSESHOE MAP. MATHEMATICAL DEFINITION OF CHAOS

The study of the evolution of sequences s of symbols S under the mapping σ is the object of *symbolic dynamics*. First, we use symbolic dynamics to show that Σ_2 contains a countable infinity of periodic orbits. This follows from the fact that for each $n = 1, 2, \ldots,$ the number of periodic points of period n is 2^n. To see this, note that periodic points under the map σ correspond to sequences consisting of repeating subsequences of length n, that is, sequences of the form

$$s = \{\cdots s_0 s_1 s_2 \cdots s_{n-1}\; s_0 s_1 s_2 \cdots s_{n-1}\; s_0 s_1 s_2 \cdots s_{n-1}\; \cdots\}.$$

(The blanks are inserted to highlight repeating subsequences.) There are two points of period 1 (i.e., fixed points, see Section 2.1.2),

$$\{\ldots 0 \quad 0 \quad 0 \ldots\} \text{ and } \{\ldots 1 \quad 1 \quad 1 \ldots\},$$

four points of period 2,

$$\{\ldots 00\ 00\ 00 \ldots\}, \{\ldots 01\ 01\ 01 \ldots\}, \{\ldots 10\ 10\ 10 \ldots\}, \{\ldots 11\ 11\ 11 \ldots\},$$

and so forth.

In addition to containing a countable infinity of periodic orbits, the dynamics in the space Σ_2 can be shown to have the following properties: its periodic points form a dense set; it has an uncountable infinity of nonperiodic orbits;

and it has a dense orbit under iteration by σ (i.e., an orbit that visits a neighborhood of any of its points)—see Appendix A4.

We now show that *the dynamics on the shift map is sensitive to initial conditions*. To do so we proceed to the following steps. First we define a metric in the space Σ_2. We recall that a metric $d(\,,)$ on a space must satisfy four axioms for every r, s, t in that space: (1) $d(r,s) \geq 0$ and $d(s,s) = 0$, (2) $d(r,s) = d(s,r)$, (3) $d(r,t) \leq d(r,s) + d(s,t)$, (4) if $r \neq s$, then $d(r,s) > 0$. For the sequence space Σ_2 the distance between two points r, s is defined as

$$d(r,s) = \sum_{i=-\infty}^{\infty} \frac{|r_i - s_i|}{2^{|i|}} \tag{3.4.1}$$

for any $r = \{\cdots r_{-k} \cdots r_{-2} r_{-1} \cdot r_0 r_1 r_2 \cdots r_k \cdots\} \in \Sigma_2$, $s = \{\cdots s_{-k} \cdots s_{-2} s_{-1} \cdot s_0 s_1 s_2 \cdots s_k \cdots\} \in \Sigma_2$. Since the numerators of the terms in the series of Eq. 3.4.1 are equal to either 0 or 1, the series is dominated by the geometric series with terms $2/2^i$ ($i = 0, 1, 2, \ldots$) and is therefore convergent. It can be seen easily that Eq. 3.4.1 satisfies each of the four axioms defining a metric (checking the third axiom involves the use of the triangle inequality, i.e., if $r, s, t \in \Sigma_2$, then $|r_i - s_i| + |s_i - t_i| \geq |r_i - t_i|$). Two points r, s are close if $r_i = s_i$ for $i = 0, \pm 1, \ldots, \pm n$, where n is large. This follows from the fact that if $r_i = s_i$ the series in Eq. 3.4.1 is dominated by the geometric series with ratio 1/2 whose first term is $2/2^{n+1}$, so $d(r,s) \leq 1/2^{n-1}$. Conversely, if $d(r,s) \leq 1/2^{n-1}$, it can be shown that $r_i = s_i$ for $i = 0, \pm 1, \ldots, \pm n$.

Second, we show that the shift map is *continuous*. Consider the sequence r and let $\epsilon > 0$. Choose n such that $1/2^{n-1} < \epsilon$. For $d(r,s) < \delta = 1/2^n$, we have $r_i = s_i$ for $i = 0, \pm 1 \ldots, \pm(n+1)$, and $d(\sigma(r), \sigma(s)) \leq 1/2^{n-1} < \epsilon$.

We are now ready to compare the behavior of point $p \in \Lambda$ under iteration by the map f with the behavior of points close to p under iteration by f. To the point p there corresponds under mapping by ϕ a sequence $\{\cdots s_{-n} \cdots s_{-1} \cdot s_0 s_1 \cdots s_n \cdots\}$ in the space Σ_2 (see Eq. 3.3.17a). For given $\epsilon > 0$, we consider an ϵ-neighborhood of p. There exists $N = N(\epsilon)$ such that the neighborhood of $\phi(p)$ corresponding to the ϵ-neighborhood of p contains sequences $s' = \{\ldots s'_{-n} \cdots s'_{-1} \cdot s'_0 s'_1 \cdots s'_n \cdots\} \in \Sigma_2$ such that $s_i = s'_i$, $|i| \leq N$. Suppose the $(N+1)$th entry in the sequence corresponding to $\phi(p)$ is 1 and the $(N+1)$th entry in a sequence s' is 0. Then, no matter how small ϵ, the condition that the Nth iterate of $\phi(p)$ and the Nth iterate of s' be close is not satisfied. In the set Λ the Nth iterate of point p is in H_1, while the Nth iterate of point p' corresponding to the sequence s' is in H_0 (Eq. 3.3.16). After N iterations, the separation between two trajectories that originally were less than ϵ apart is at least $1 - 2\mu$.

Recall that, owing to topological conjugacy (Appendix A3), the dynamical properties of the map σ in Σ_2 (Appendix A4) carry over to the map f in Λ. It follows that, like the map σ in Σ_2, the Smale horseshoe map has a Cantor

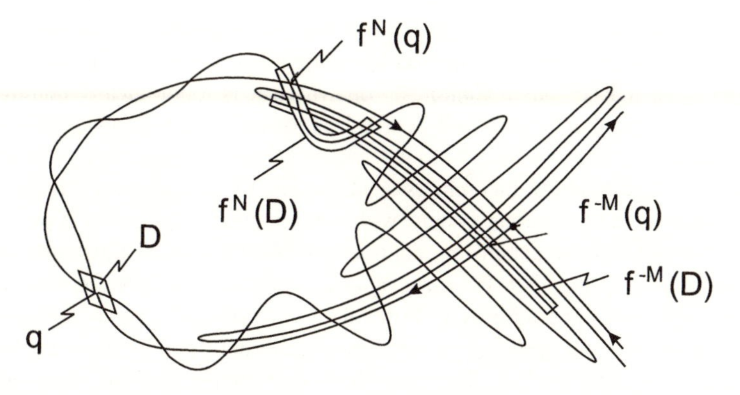

Figure 3.9. Illustration of intersection in a horseshoelike configuration occurring in a homoclinic tangle (after Arrowsmith and Place, 1990).

set Λ with the following properties:

1. Λ contains a countable infinity of periodic orbits with arbitrarily high period. These orbits are dense in Λ.
2. Λ contains an uncountable infinity of nonperiodic orbits.
3. Λ contains a dense orbit under iteration by f (i.e., the map is topologically transitive).

At this point we can define chaos mathematically. Let V be a set. A map $F: V \to V$ is said to be chaotic if it is sensitive to initial conditions and has the three properties listed above. By virtue of this definition it follows that the Smale horseshoe map is chaotic.

3.5 SMALE-BIRKHOFF THEOREM. MELNIKOV NECESSARY CONDITION FOR CHAOS. TRANSIENT AND STEADY-STATE CHAOS

Consider a map $f: \mathbf{R}^2 \to \mathbf{R}^2$, assumed to be a diffeomorphism[2] and to possess a fixed hyperbolic point whose stable and unstable manifolds intersect transversely (Fig. 3.9). The parallelogram D in Fig. 3.9 contains the homoclinic point q and has sides parallel to the directions of the stable and unstable manifolds at that point. It maps to the areas denoted by $f^N(D)$ and $f^{-M}(D)$, respectively, under N forward iterations and M reverse iterations of the map f. The mapping entails expansion, contraction, folding, and intersections in a horseshoe configuration, similar to those occurring in the Smale horseshoe

[2]A C^r diffeomorphism is a homeomorphism whose continuous inverse f^{-1} has derivatives of order $r \geq 1$.

map. This similarity can be used to prove the Smale-Birkhoff theorem, which states that, if the Melnikov function has simple zeros, there exists an integer $n \geq 1$ such that f^n has a Cantor set on which it is topologically conjugate to a shift map. For a long and detailed exposition see Wiggins (1990, pp. 443–482).

The Smale-Birkhoff theorem is a statement of the necessary condition for chaos for a continuous system with a homoclinic tangle. For a *periodically excited flow* the expansion, contraction, and folding are illustrated for a Poincaré map in Fig. 3.9. For a *quasiperiodically excited flow* a geometrical object topologically conjugate to Fig. 3.9 can be created by projecting on the same plane $\{x_1, x_2\}$ appropriate portions of different phase space slices such as, e.g., those of Fig. 2.11. The horseshoes that intersect in this plane originate from phase space slices taken at equal time intervals. Beigie et al. (1991) referred to them as *traveling horseshoe sequences* and showed that they can be used just like the Poincaré map to prove that the necessary condition for the occurrence of chaos in a quasiperiodically excited system is that the system's Melnikov function have simple zeros.

Chaos in a dissipative dynamical system implies the existence of a chaotic attractor and/or a chaotic nonattracting set. A *chaotic attractor* (Section 3.1.3) is a bounded set defined by steady-state chaotic motion in the phase space. By virtue of this definition an attractor cannot be defined in the nonautonomous phase space $\{x_1, x_2, t\}$, since in that space the steady-state motion is unbounded. This difficulty is removed if the system is suspended in an autonomous phase space (see Sections 2.7.1 and 2.7.3). Fractal boundaries (also referred to as blurred boundaries) of a basin of attraction, that is, basin boundaries with fractal dimensions (see, e.g., Ott, 1997 or Moon and Li, 1985) are an example of a nonattracting chaotic set.

Consider a system with parameter γ. Let γ_{tg} denote the parameter value for which the stable and unstable manifolds are tangent, that is, for which the Melnikov function has double zeros, and assume that for $\gamma \leq \gamma_{tg}$ the Melnikov condition for chaos is not satisfied. Since the manifolds do not intersect transversely, the system has no steady-state or transient chaotic motions, and the boundaries of its basin of attraction are smooth. There exists a parameter value γ_{chaos} such that steady state chaotic motions occur for $\gamma \geq \gamma_{chaos}$.[3] Numerical experiments indicate that, depending upon initial conditions, for $\gamma_{tg} < \gamma < \gamma_{chaos}$ the system may have a nonattracting chaotic set, the size and configuration of which depend on γ, as well as transient motions (motions occurring before the motion on the chaotic attractor, i.e., the steady-state motion, is reached) that are sensitive to initial conditions and are referred to as chaotic transients; see Example 3.5.1.

[3]The chaotic attractor can coexist with other attractors, either chaotic or nonchaotic, and with a nonattracting chaotic set.

Example 3.5.1 *Regular and chaotic motions in a harmonically excited standard Duffing–Holmes oscillator*

We consider the harmonically excited standard Duffing–Holmes oscillator (Eqs. 2.5.14 in which $a = b = 1$) with parameters $\beta = 0.25$, $\omega = 1.0$. From Eqs. 2.5.17 and 2.5.19 it follows that the Melnikov condition for chaos is $\gamma > \gamma_{tg} = 4\beta/[3\pi(2)^{1/2}\omega\,\mathrm{sech}(\pi\omega/2)] = 0.188255$.

For $\gamma = 0$, the energy of the motion is gradually dissipated and, depending upon initial conditions, the system settles on one of two attractors. Each attractor is a fixed point at the bottom of one of the two potential wells shown in Fig. 2.1a. Figure 3.4 shows a phase plane diagram for this case, including the two attractors, their respective basins of attraction (shaded for the right attractor and blank for the left attractor), the basin boundaries, and two trajectories moving each toward an attractor. Note that the basin boundaries are smooth.

For $\gamma = 0.10$ the system has two periodic attractors in the autonomous phase space. In a three-dimensional phase space $\{x_1, x_2, t\}$ the steady-state motion occurs on a spiral. The spiral intersects a Poincaré plane of section at a point. Figure 3.10a shows that the intersections of the Poincaré plane of section with the stable and unstable manifolds are nonintersecting curves. A displacement time history, which includes a transient, nonchaotic portion up to time $t \approx 30$ and a steady-state periodic motion thereafter, is shown in Fig. 3.10b. The motion in this case is not sensitive to initial conditions.

For $\gamma = 0.20$ the stable and unstable manifolds intersect transversely (Fig. 3.11a), but no chaotic attractor is observed. Rather, the system has two periodic attractors similar to but, because the excitation amplitude is greater, somewhat larger than those of Fig. 3.10a. The time history of Fig. 3.11b, for which the initial conditions are $(0.1, 0.1)$, features a transient that, unlike the transient of Fig. 3.10b, is chaotic and therefore sensitive to initial conditions, as can be verified numerically. Transient motions corresponding to other initial conditions (e.g., $(0, 0)$) are not chaotic and the corresponding motions are therefore not sensitive to initial conditions.

The chaotic or nonchaotic character of transients corresponding to any given set of initial conditions depends upon the configuration of the basins of attraction and their boundaries. According as the Melnikov function does not or does have simple zeros, basin boundaries are smooth or may be fractal (blurred). An increased excitation causes an *erosion* of—an increased prevalence of fractal regions in—the basins of attraction (Moon and Li, 1985; Moon, 1992).

For $\gamma = 0.40$, Figs. 3.12a,b,c show a Poincaré section through the intersecting manifolds, a Poincaré map of the chaotic attractor, and a chaotic displacement time history, respectively. Melnikov theory requires that the perturbation be sufficiently small. Figures 3.11 and 3.12 show that the Melnikov

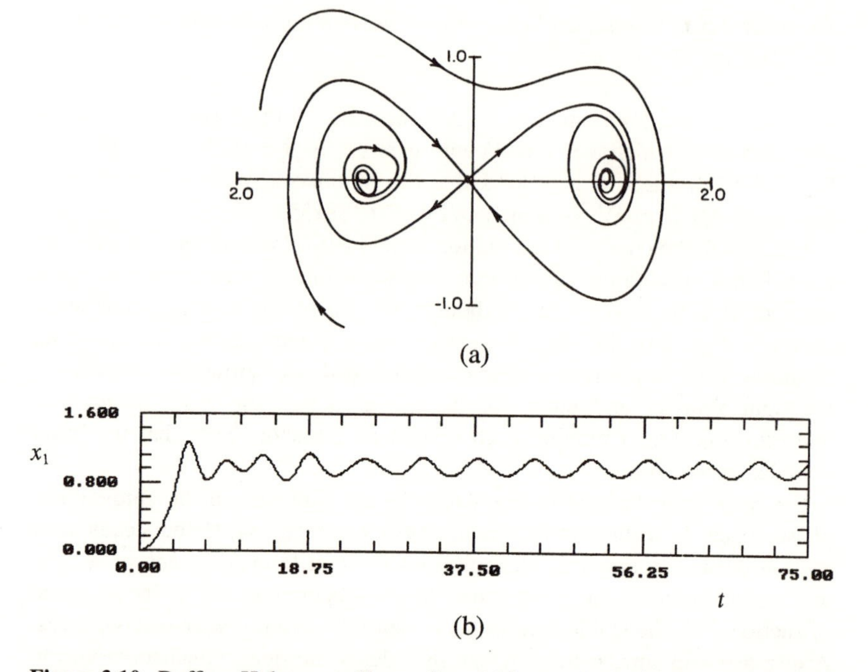

Figure 3.10. Duffing–Holmes oscillator, $\beta = 0.25$, $\gamma = 0.10$: (a) Poincaré section showing nonintersecting stable and unstable manifolds (from Guckenheimer and Holmes, 1986); (b) displacement time history.

necessary condition for chaos is applicable in practice even for relatively large perturbations. For an experimental confirmation see Section 3.7.

3.6 CHAOTIC DYNAMICS IN PLANAR SYSTEMS WITH A SLOWLY VARYING PARAMETER

As indicated in Section 2.8, if the Melnikov function of a slowly varying planar system has a simple zero at some time t_0, for sufficiently small ϵ the stable and unstable manifolds of the perturbed system intersect transversely. Conversely, if $M(t_0) \neq 0$ for all t_0, then the stable and unstable manifolds do not intersect. If the excitation is periodic in time, then it is possible to construct a Poincaré map in the phase space $\{x, y, z\}$ that is topologically equivalent to a three-dimensional Smale horseshoe map. The map contracts the x and y directions and stretches the z direction. Following a technique similar to that used for the planar case the map can be shown to be topologically equivalent to a shift map acting on a space of bi-infinite sequences of 0's and 1's. Therefore, for the system 2.8.1 to be chaotic it is necessary that the Melnikov function have simple zeros. For details, and a version of the

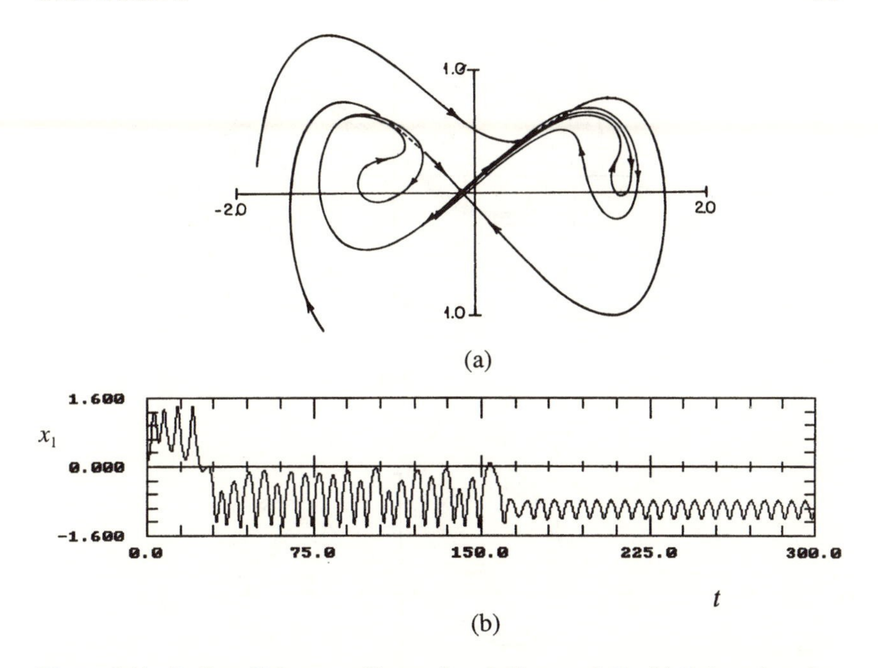

Figure 3.11. Duffing–Holmes oscillator, $\beta = 0.25$, $\gamma = 0.20$: (a) Poincaré section through intersecting stable and unstable manifolds (from Guckenheimer and Holmes, 1986); (b) displacement time history. The motion begins with a chaotic transient; however, the steady-state motion is periodic. The Melnikov necessary condition for chaos is satisfied, that is, the stable and unstable manifolds intersect transversely but, as this example shows, while this is necessary, it is not sufficient for steady-state chaotic motion to occur.

Smale-Birkhoff theorem applicable to maps with homoclinic points resulting from the intersection of one-dimensional with two-dimensional invariant manifolds, see Wiggins and Shaw (1988).

As was pointed out in Section 2.8.3, unlike planar systems with no slowly varying parameter, some slowly varying planar systems experience escapes by mechanisms other than chaotic transport, that is, nonchaotic escapes can occur independently of the Melnikov function's behavior. For an example see Whalen (1996).

3.7 CHAOS IN AN EXPERIMENTAL SYSTEM: THE STOKER COLUMN

In this section we describe a modified Stoker column (Stoker, 1950), whose harmonically forced motions demonstrate the existence of chaos. The col-

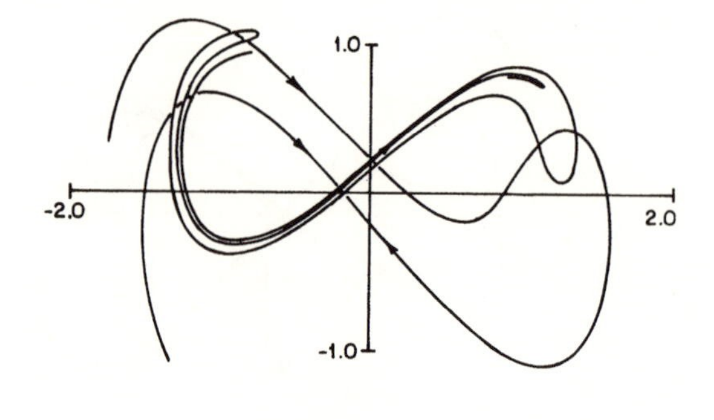

(a)

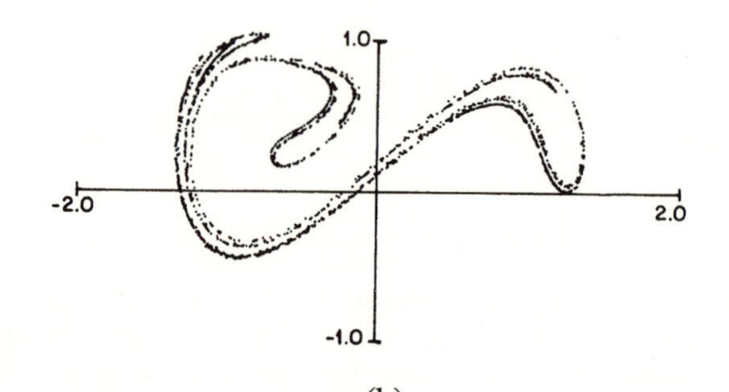

(b)

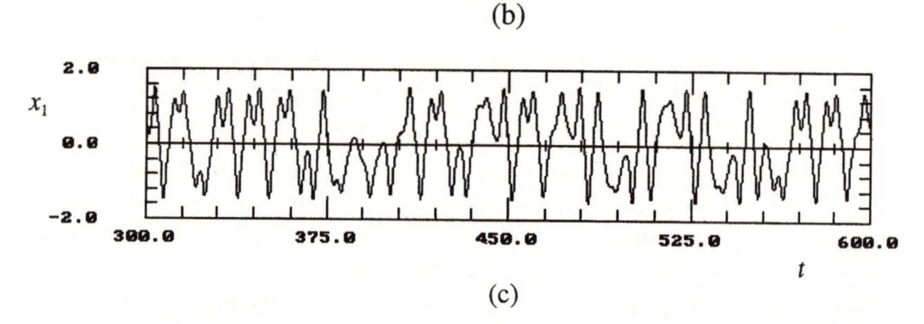

(c)

Figure 3.12. Duffing–Holmes oscillator, $\beta = 0.25$, $\gamma = 0.40$: (a) Poincaré section through intersecting stable and unstable manifolds; (b) Poincaré map of the chaotic attractor (from Guckenheimer and Holmes, 1986); (c) displacement time history. The chaotic attractor, and its Poincaré map, are fractal objects.

umn is described schematically in Fig. 3.13, and the constructed device is shown in Fig. 3.14. The system has three concentrated masses: a mass m at its midpoint, a mass m_1 at the top end, and a mass m_2 at the bottom end. Torsional springs at the midpoint supply the total concentrated bending stiffness K_1. The column, whose length is 2ℓ ($\ell = 200$ mm), is tied at its

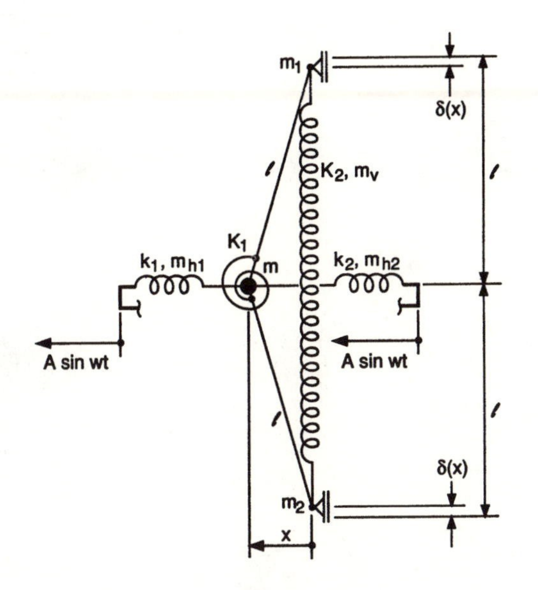

Figure 3.13. Schematic of forced modified Stoker column (after Cook and Simiu, 1991).

ends by vertical springs with total mass m_v and total stiffness K_2. The column ends and the masses attached to them are free to slide vertically but are constrained in the horizontal direction. The mass m is free to slide horizontally but is constrained in the vertical direction. The mass m is tied by two springs (with masses m_{h1} and m_{h2} and stiffnesses k_1 and k_2) to two rigid arms forced by a motor to move harmonically (with amplitude A and circular frequency ω) and in phase in the horizontal direction (Cook and Simiu, 1991).

Figure 3.15 shows a time history of the displacement. The motion is seen to be highly irregular, even though the excitation is deterministic. It is seen in Fig. 3.16 that the spectral density associated with the irregular displacement is broadband.[4] For a deterministic system this is a hallmark of chaos, which implies an uncountable infinity of nonperiodic orbits (Section 3.4). Had the response been deterministic the spectral density would have consisted of discrete spikes. At higher frequencies the continuous portion of the spectrum is at least two orders of magnitude lower than at lower frequencies and may

[4]The spectral density function is a measure of the amplitudes of the elemental harmonic components into which a stochastic process may be resolved—see Section 4.2 for details.

Figure 3.14. Constructed modified Stoker column device (from Cook and Simiu, 1991).

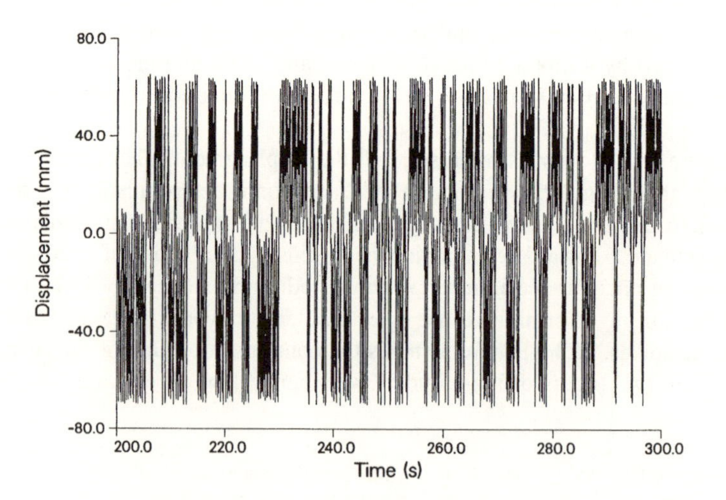

Figure 3.15. Time history of displacement.

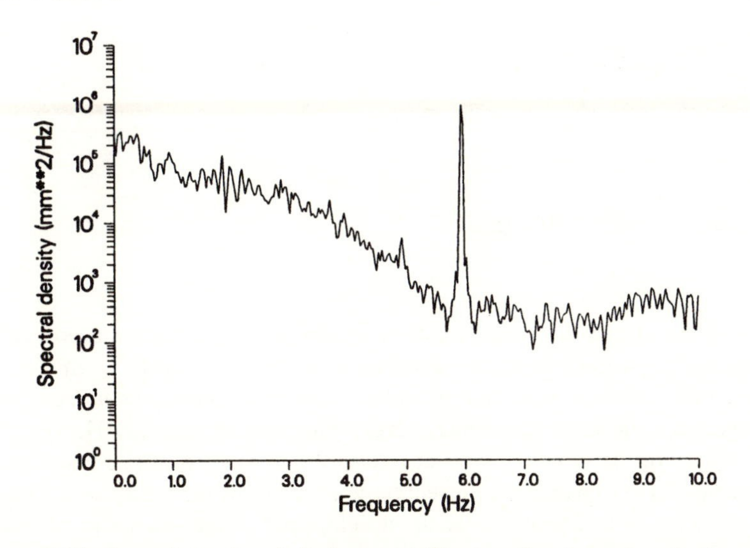

Figure 3.16. Spectral density of displacement. The broadband spectrum is due predominantly to the chaotic motion, but instrumental and numerical noise, also present in spectra of periodic motions, account for a small part of the total spectral content.

be attributed to instrumental and/or numerical noise. Figure 3.17 shows a Poincaré map of the chaotic attractor. Its estimated fractal dimension was 1.3, that is, intermediate between the dimension of a line and the dimension of a surface.

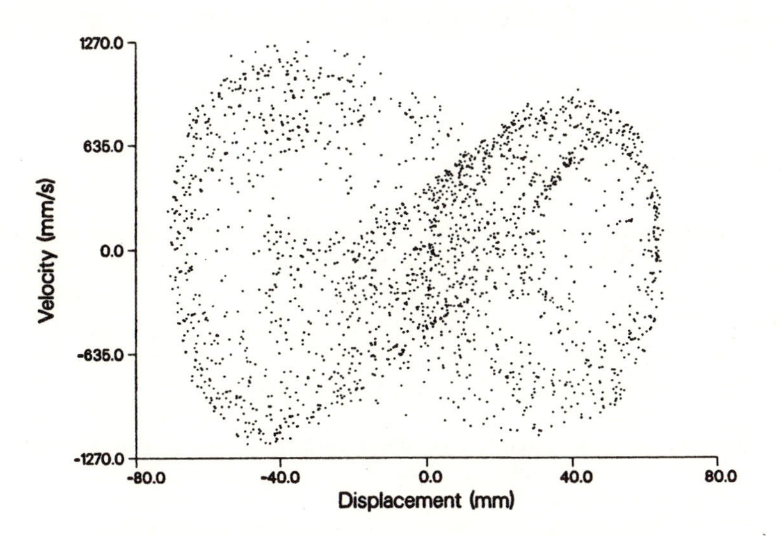

Figure 3.17. Poincaré plot of steady-state chaotic motion.

Chapter Four

Stochastic Processes

Probability theory is a mathematical model for the description and interpretation of phenomena represented by variables, referred to as *random* or *stochastic variables*, that show statistical variability.[1] *Stochastic processes* are defined as collections of infinitely many functions of time $\{y(t)\}$, such that the values of the functions $y(t)$ at any specified time constitute a stochastic variable.[2] In this chapter we present basic elements of the theory of stochastic processes used in Chapter 5 for the development of the stochastic Melnikov approach. Basic elements of probability theory are reviewed in Appendix A5.

Section 4.1 presents basic definitions and results of the theory of stochastic processes, including definitions of the spectral density function, the autocovariance function, and the cross-covariance function. Section 4.2 is devoted to approximate representations of stochastic processes as finite sums of harmonic functions with random parameters. Such representations are needed for the extension of the Melnikov approach to stochastic systems. Section 4.3 includes the derivation of the expression for the spectral density of the output of a linear filter with a stochastic input. This expression is used in Chapter 5 for the calculation of the spectral density of the stochastic counterpart of the Melnikov function.

4.1 SPECTRAL DENSITY, AUTOCOVARIANCE, CROSS-COVARIANCE

In this section we introduce basic material on stochastic processes and their description. Following a review of basic definitions and of requisite elements of harmonic analysis, we define two mutually dependent descriptors of a stochastic process, the spectral density function and the autocovariance function. We then define a descriptor of the relationship between two distinct stochastic processes, the cross-covariance function.

[1]The word "*stochastic*" means "connected with random experiments and probability," and is derived from the Greek στοχαζομαι, meaning "to aim at, seek after, surmise." The terms "random" and "stochastic" may be used interchangeably.

[2]A broader definition that will not be used in this text entails dependence on space variables as well.

4.1.1 Basic Definitions

Consider processes whose outcomes form a collection—an *ensemble*—of infinitely many functions of time $\{y(t)\}$. A member of the ensemble, denoted by $y_i(t)$ $(i = 1, 2, \ldots)$ is referred to as a *sample function*, a *path*, a *signal*, or a *realization* of the ensemble. The process is called a stochastic process if the values of the realizations at any specified time constitute a stochastic variable. A stochastic process is called *stationary* if its statistical properties are not dependent upon the choice of the time origin, that is, if "whatever started to happen at some time could equally have started at any other time." Unless otherwise noted we will consider in this book only stationary stochastic processes. By definition, a stationary sample path is assumed to extend over the entire time domain. However, this is a mere idealization: physical processes are in fact of finite, though possibly very long, duration. We note that noise—a stochastic process—commonly denotes a small stochastic addition to a deterministic signal; nevertheless, the terms "noise" and "stochastic process" may be used interchangeably, and we do so in this book.

The probability distribution of $y(t)$ at any fixed time t is the *marginal distribution* of the stochastic process. It follows from the definition of stationarity that in statistically stationary processes the marginal distribution is independent of time. If the distributions of the random vectors $(y(t_1), \ldots, y(t_n))$ for any integer $n \geq 1$ and times $t_1 < t_2 < \cdots < t_n$ are Gaussian (see Appendix A5, Section A5.6) the process is referred to as *Gaussian*. Otherwise the process is called *non-Gaussian*.

The *ensemble average*, or the *expectation*, of a stationary process is the average of the values of the member functions at any fixed time. A stationary stochastic process is said to be *ergodic* if its ensemble and temporal averages of functions of $y(t)$ coincide. It follows that in an ergodic process every sample function is typical of the entire ensemble. Unless otherwise noted, the stochastic processes we are concerned with are assumed to be stationary and ergodic. Realizations of a stationary ergodic process are represented in Fig. 4.1.

A full description of a stochastic process would require knowledge of the joint distribution of the variables $y_t(t_j)$ for fixed i (i.e., for a specified realization) and for all times $t_1, \ldots, t_n$. In general such knowledge is not available. However, a useful description of the process is provided by the marginal distribution and the *spectral density function*[3] or its inverse Fourier transform, the *autocovariance function*, which contain information on how a process with given marginal distribution evolves in time. We now proceed to defining these functions.

[3]The spectral density function is also referred to as the spectral density or the spectrum.

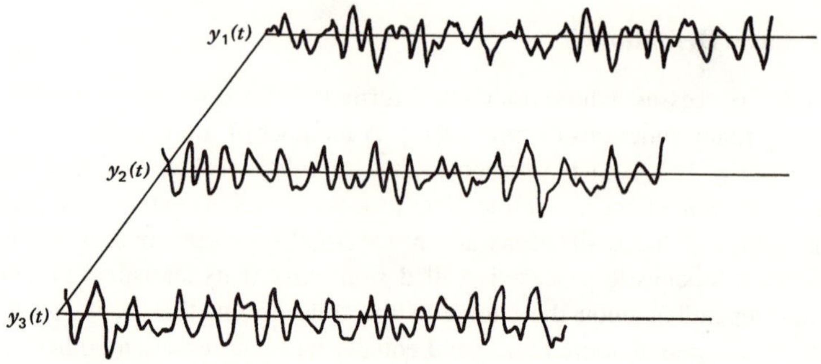

Figure 4.1. Realizations of a stationary ergodic process.

4.1.2 Fourier Series and Fourier Integrals

The idea behind the mathematical construct known as spectral density is that a realization of a stochastic process may be viewed as a sum of harmonic contributions. In this section we review representations of periodic and nonperiodic deterministic functions as sums of harmonic contributions, that is, Fourier series and Fourier integrals. The definition of the spectral density will be based on these representations.

4.1.2.1 Periodic Functions: Fourier Series

Consider a periodic function $x(t)$ with period T. The Fourier series expansion of $x(t)$ can be written as

$$x(t) = C_0 + \sum_{k=1}^{\infty} C_k \cos(k\omega_1 t - \phi_k). \tag{4.1.1a}$$

In Eq. 4.1.1a, $\omega_1 = 2\pi/T$ is the *fundamental circular frequency,* or simply the *fundamental frequency*, and

$$C_0 = \frac{1}{T}\int_{-T/2}^{T/2} x(t)\,dt, \tag{4.1.1b}$$

$$C_k = (A_k^2 + B_k^2)^{1/2}, \tag{4.1.1c}$$

$$\phi_k = \tan^{-1}(B_k/A_k), \tag{4.1.1d}$$

$$A_k = \frac{2}{T}\int_{-T/2}^{T/2} x(t)\cos k\omega_1 t\,dt, \tag{4.1.1e}$$

$$B_k = \frac{2}{T}\int_{-T/2}^{T/2} x(t)\sin k\omega_1 t\,dt. \tag{4.1.1f}$$

To verify Eqs. 4.1.1, note that

$$\cos(k\omega_1 t - \phi_k) = \cos kw_1 t \cos\phi_k + \sin k\omega_1 t \sin\phi_k, \qquad (4.1.1g)$$

so, by virtue of Eqs. 4.1.1c,d,

$$x(t) = C_0 + \sum_{k=1}^{\infty}(A_k \cos k\omega_1 t + B_k \sin k\omega_1 t). \qquad (4.1.1h)$$

Equations 4.1.1b,e,f are verified by substituting Eq. 4.1.1h into their respective right-hand sides. Equation 4.1.1h, and therefore Eq. 4.1.1a, can be written in complex form as

$$\boldsymbol{x}(t) = \sum_{k=-\infty}^{\infty} \boldsymbol{C}_k \exp(jk\omega_1 t) \qquad (4.1.2a)$$

$$\equiv x(t), \qquad (4.1.2b)$$

$$\boldsymbol{C}_k = \frac{1}{T}\int_{-T/2}^{T/2} x(t)\exp(-jk\omega_1 t)\,dt, \qquad (4.1.2c)$$

where $j = \sqrt{-1}$, and the real part of the complex function $\boldsymbol{x}(t)$ is $x(t)$. As indicated by Eq. 4.1.2b, the imaginary part of $\boldsymbol{x}(t)$ vanishes, since it consists of the sum

$$\sum_{k=-\infty}^{\infty}\left\{\left[\frac{1}{T}\int_{-T/2}^{T/2} x(t)\cos(k\omega_1 t)dt\right]\sin k\omega_1 t - \left[\frac{1}{T}\int_{-T/2}^{T/2} x(t)\sin(k\omega_1 t)dt\right]\cos(k\omega_1 t)\right\}, \qquad (4.1.2d)$$

in which each term corresponding to k is equal and of opposite sign to its counterpart for $-k$.

4.1.2.2 Nonperiodic Functions: Fourier Integrals and Fourier Transforms

A function $y(t)$ that is nonperiodic may be viewed as periodic with infinite period. We assume that $y(t)$ is piecewise differentiable in every finite interval, and that the integral

$$\int_{-\infty}^{\infty} |y(t)|dt \qquad (4.1.3)$$

exists.

Assuming that the fundamental frequency ω_1 is small and denoting it by $\Delta\omega$, we combine Eqs. 4.1.2, in which we substitute $y(t)$ for $x(t)$, as follows:

$$\mathbf{y}(t) = \sum_{k=-\infty}^{\infty} \left[(\Delta\omega/2\pi) \int_{-T/2}^{T/2} y(t) \exp(-j\omega_k t) dt \right] \exp(j\omega_k t). \qquad (4.1.4)$$

(Recall that $y(t)$ is the real part of the complex function $\mathbf{y}(t)$.) We now let $\Delta\omega$ become the infinitesimal quantity $d\omega$, so that $T \to \infty$ and $\omega_k \equiv k\Delta\omega = \omega$. We obtain the result, analogous to Eq. 4.1.2,

$$\mathbf{y}(t) = \int_{-\infty}^{\infty} \frac{d\omega}{2\pi} \left[\int_{-\infty}^{\infty} y(t) \exp(-j\omega t)\, dt \right] \exp(j\omega t)$$

or

$$\mathbf{y}(t) = \frac{1}{2\pi} \int_{-\infty}^{\infty} \mathbf{C}(\omega) \exp(j\omega t)\, d\omega \qquad (4.1.5a)$$

$$\equiv y(t), \qquad (4.1.5b)$$

$$\mathbf{C}(\omega) = \int_{-\infty}^{\infty} y(t) \exp(-j\omega t)\, dt. \qquad (4.1.6)$$

Equation 4.1.5b follows from the same observation that led to Eq. 4.1.2b. The functions $y(t)$ and $\mathbf{C}(\omega)$ are referred to as the *inverse Fourier transform* of $\mathbf{C}(\omega)$ and the *Fourier transform* of $y(t)$, respectively, and are said to form a *Fourier transform pair.*

4.1.2.3 Time Derivative of $y(t)$

In subsequent applications we need to use expressions containing the derivative of $y(t)$. The latter is obtained by differentiating Eq. 4.1.5 with respect to time:

$$\dot{y}(t) = \frac{j}{2\pi} \int_{-\infty}^{\infty} \omega \mathbf{C}(\omega) \exp(j\omega t)\, d\omega. \qquad (4.1.7)$$

4.1.3 Parseval's Equality

In this section we introduce Parseval's equality as a preliminary step toward defining the spectral density function of a stochastic process.

4.1.3.1 Parseval's Equality for Periodic Functions

First we consider a periodic function $x(t)$ with zero mean and period T (Eq. 4.1.1). The variance of $x(t)$ is

$$\sigma_x^2 = \frac{1}{T}\int_{-T/2}^{T/2} x^2(t)\,dt. \tag{4.1.8}$$

The substitution of Eq. 4.1.1a into Eq. 4.1.8 yields

$$\sigma_x^2 = \sum_{k=0}^{\infty} \Psi_k \tag{4.1.9}$$

where $\Psi_0 = C_0^2 = 0$ (by virtue of the assumption that $x(t)$ has zero mean and of Eq. 4.1.1b), and $\Psi_k = (1/2)C_k^2$ $(k = 1, 2, \ldots)$. Equation 4.1.9 is referred to as Parseval's equality. *The quantity Ψ_k is the contribution to the variance of $x(t)$ of the harmonic component with frequency $k\omega_1$.*

4.1.3.2 Parseval's Equality for Nonperiodic Functions

Consider now a nonperiodic function $y(t)$ with zero mean for which Eq. 4.1.3 is satisfied. By Eqs. 4.1.5 and 4.1.6 we have

$$\int_{-\infty}^{\infty} y^2(t)\,dt = \frac{1}{2\pi}\int_{-\infty}^{\infty} y(t)\left[\int_{-\infty}^{\infty} \mathbf{C}(\omega)\exp(j\omega t)\,d\omega\right]dt$$

$$= \frac{1}{2\pi}\left[\int_{-\infty}^{\infty} \mathbf{C}(\omega)\int_{-\infty}^{\infty} y(t)\exp(j\omega t)\,d\omega\,dt\right]$$

$$= \frac{1}{2\pi}\left[\int_{-\infty}^{\infty} \mathbf{C}(\omega)\mathbf{C}^*(\omega)\,d\omega\right] \tag{4.1.10a}$$

$$= \frac{1}{2\pi}\int_{-\infty}^{\infty} C^2(\omega)\,d\omega. \tag{4.1.10b}$$

In Eqs. 4.1.10a and 4.1.10b the symbols $\mathbf{C}^*(\omega)$ and $C(\omega)$ denote, respectively, the complex conjugate of $\mathbf{C}(\omega)$ and the real part of $\mathbf{C}(\omega)$. Equations 4.1.10 are the form taken by Parseval's equality for a nonperiodic function.

4.1.4 The Spectral Density Function

4.1.4.1 Definition of the Spectral Density Function

We now seek relations similar to Eq. 4.1.10 for a stochastic process with zero mean. The spectral density is defined as the counterpart for the stochastic process of the quantities Ψ_k (Eq. 4.1.9).

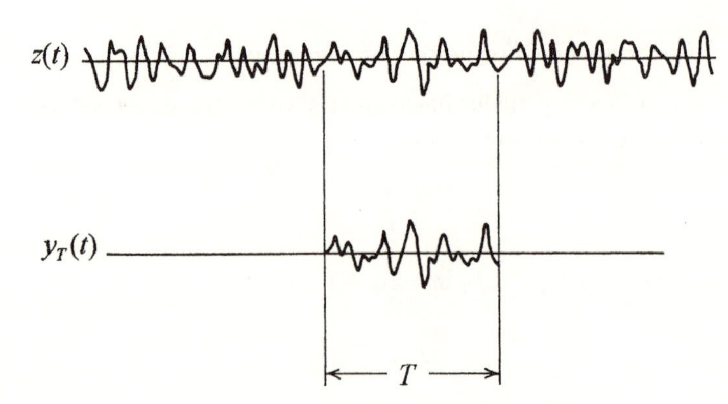

Figure 4.2. Definition of function $y_T(t)$.

Consider a stochastic signal $z(t)$ with zero mean. Because it does not satisfy the condition 4.1.3 the function $z(t)$ does not possess a Fourier transform. An auxiliary function $y_T(t)$ is therefore defined as follows (Fig. 4.2):

$$y_T(t) = z(t)\left(-\frac{T}{2} < t < \frac{T}{2}\right) \tag{4.1.11a}$$

$$y_T(t) = 0 \qquad \text{elsewhere.} \tag{4.1.11b}$$

The function $y_T(t)$ so defined is nonperiodic, satisfies condition 4.1.3, and therefore has a Fourier integral. From the definition of $y_T(t)$ it follows that

$$\lim_{T\to\infty} y_T(t) = z(t). \tag{4.1.12}$$

By virtue of Eqs. 4.1.11 and 4.1.10b the variance of $y_T(t)$ is

$$\sigma_{y_T}^2 = \frac{1}{T}\int_{-T/2}^{T/2} \left[y_T(t)\right]^2 dt \tag{4.1.13a}$$

$$= \frac{1}{T}\int_{-\infty}^{\infty} \left[y_T(t)\right]^2 dt \tag{4.1.13b}$$

$$= \frac{1}{2\pi T}\int_{-\infty}^{\infty} C_T^2(\omega)\, d\omega. \tag{4.1.13c}$$

The variance of the function $z(t)$ is then

$$\sigma_z^2 = \lim_{T\to\infty} \sigma_{y_T}^2$$

$$= \lim_{T\to\infty} \frac{1}{2\pi T}\int_{-\infty}^{\infty} C_T^2(\omega)\, d\omega. \tag{4.1.14}$$

We now introduce the notation

$$\Psi_z(\omega) = \lim_{T\to\infty}\left(\frac{1}{T}\right)C_T^2(\omega). \tag{4.1.15}$$

Equation 4.1.14 can be written

$$\sigma_z^2 = \frac{1}{2\pi}\int_{-\infty}^{\infty} \Psi_z(\omega)\, d\omega. \tag{4.1.16}$$

The function $\Psi_z(\omega)$ is defined as the spectral density function of the stochastic process $z(t)$. The area under the spectral density curve $\Psi_z(\omega)$ is the variance σ_z^2 of $z(t)$. *The quantity $\Psi_z(\omega)\, d\omega$ is the elemental contribution to the variance of $z(t)$ of the harmonic component with frequency ω.*

4.1.4.2 One-Sided and Two-Sided Spectral Density Functions

In Eq. 4.1.16 the frequencies ω are defined over the domain $-\infty < \omega < \infty$, and the spectrum $\Psi_z(\omega)$ is called *two sided.* In some applications it is desirable to define the spectral density over the frequency domain $0 < \omega < \infty$. The spectral density is then referred to as *one sided*, and is defined as

$$\Psi_z^{\text{o.s.}}(\omega) = \Psi_z(\omega) + \Psi_z(-\omega). \tag{4.1.17a}$$

It follows from Eq. 4.1.16 that

$$\sigma_z^2 = \frac{1}{2\pi}\int_0^{\infty} \Psi_z^{\text{o.s.}}(\omega)\, d\omega. \tag{4.1.17b}$$

Henceforth the superscript "o.s." will be omitted whenever it is indicated in the text that the spectral density function is one sided. To illustrate the correspondence between a path $z(t)$ and its one-sided spectral density, we show in Fig. 4.3 the time history of $z(t)$ for the case where $\Psi_z^{\text{o.s.}}(\omega)$ has two peaks at the frequencies ω_1, ω_2.

4.1.4.3 Spectral Density of Time Derivative of $z(t)$

From Eq. 4.1.7, following steps similar to those that led from Eq. 4.1.10 to Eq. 4.1.16, we obtain the expression for the spectral density of the first derivative $\dot{z}(t)$ of a stochastic process $z(t)$:

$$\Psi_{\dot{z}}(\omega) = \omega^2 \Psi_z(\omega). \tag{4.1.18}$$

Equation 4.1.18 is used in Appendix A6 for determining the joint probability distribution of z and $\dot{z}$, which in turn is necessary for obtaining the expression for the rate of zero upcrossing of the process $z(t)$.

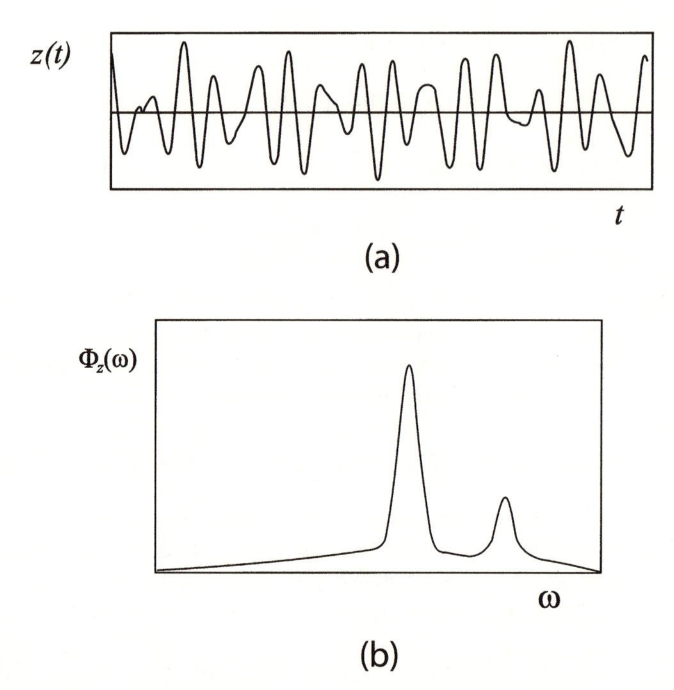

Figure 4.3. (a) Time history of a path $z(t)$ for a stochastic process whose one-sided spectral density $\Psi_z(\omega)$, shown in (b), has peaks at the frequencies ω_1 and ω_2. The path exhibits fluctuations dominated by components with average frequencies ω_1 and ω_2.

4.1.5 Autocovariance Function: Definition and Physical Interpretation

In this section we define the autocovariance function of a stationary stochastic variable $z(t)$ with zero mean. We show that the autocovariance function forms with the spectral density function a Fourier transform pair. We then provide a physical interpretation of the autocovariance function and its relation to the spectral density function.

4.1.5.1 Autocovariance Function

The autocovariance function $R_z(\tau)$ of a stationary stochastic process $z(t)$ with zero mean is defined by the expression

$$R_z(\tau) = \lim_{T\to\infty} \frac{1}{T} \int_{-T/2}^{T/2} z(t)z(t+\tau)\, dt. \tag{4.1.19}$$

The following derivations lead to the result that the spectral density function is the Fourier transform of the autocovariance function, that is,

$$\Psi_z(\omega) = \int_{-\infty}^{\infty} R_z(\tau) \exp(-j\omega\tau)\, d\tau. \tag{4.1.20}$$

Equations 4.1.20 and 4.1.5, 4.1.6 imply

$$R_z(\tau) = \frac{1}{2\pi} \int_{-\infty}^{\infty} \Psi_z(\omega) \exp(j\omega\tau)\, d\omega, \tag{4.1.21}$$

that is, the autocovariance function is the inverse Fourier transform of the spectral density function, and forms with the spectral density function a Fourier transform pair.

To derive Eq. 4.1.20 we note that Eq. 4.1.6 yields

$$\begin{aligned}\frac{1}{T} C^2(\omega) &= \frac{1}{T} C(\omega)\, C^*(\omega)\\ &= \frac{1}{T}\left[\int_{-\infty}^{\infty} y(t_1) \exp(-j\omega\, t_1\, dt_1) \int_{-\infty}^{\infty} y(t_2) \exp(j\omega t_2)\, dt_2\right]\\ &= \frac{1}{T} \int_{-\infty}^{\infty} \int_{-\infty}^{\infty} y(t_1) y(t_2) \exp[j\omega(t_2 - t_1)]\, dt_1\, dt_2. \end{aligned} \tag{4.1.22}$$

We obtain Eq. 4.1.20 by taking the limit of Eq. 4.1.22 as $T \to \infty$, using Eqs. 4.1.11, 4.1.15, 4.1.19, and denoting $\tau = t_2 - t_1$.

We have

$$R_z(\tau) = R_z(-\tau). \tag{4.1.23}$$

This is a consequence of Eq. 4.1.19 and the stationarity of the function $z(t)$, by virtue of which the integrand of Eq. 4.1.19 can be written as $z(t - \tau)z(t)$. It follows from Eqs. 4.1.20 and 4.1.23 that the left-hand side of Eq. 4.1.21 is a real function.

By Eq. 4.1.19, the autocovariance function represents the time average of the product $z_1(t)z_2(t)$ of the functions $z_1(t) \equiv z(t)$ and $z_2(t) \equiv z(t - \tau_0)$ (Fig. 4.4). Since we assume that the process is ergodic, $R_z(\tau)$ is also equal to the ensemble average of the products $z(t)z(t + \tau)$ for any fixed time t, that is,

$$R_z(\tau) = E[z(t)z(t + \tau)]. \tag{4.1.24}$$

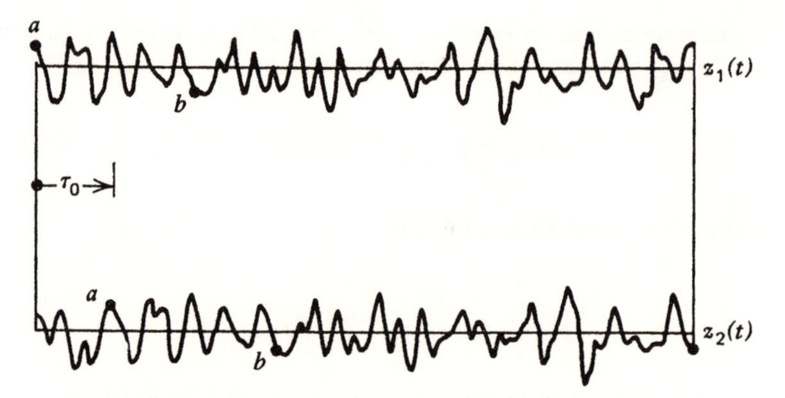

Figure 4.4. Functions $z_1(t) \equiv z(t)$ and $z_2(t) \equiv z(t - \tau_0)$.

4.1.5.2 Physical Interpretation of the Autocovariance Function

For $\tau = 0$, Eqs. 4.1.21 and 4.1.16 yield

$$R_z(0) = \sigma_z^2. \qquad (4.1.25)$$

Since $\Psi_z(\omega) > 0$ (Eq. 4.1.15) and, for $\tau = 0$, $|\exp(j\omega\tau)| \equiv 1$, it follows from Eq. 4.1.21 that, for $\tau > 0$, $R_z(\tau) < R_z(0)$, or

$$R_z(\tau) < \sigma_z^2. \qquad (4.1.26)$$

The extent to which the autocovariance function is smaller than the variance depends on the time lag τ *and* the spectral density of the process. Consider, for example, a process with relatively sluggish fluctuations, that is, one whose dominant period is relatively large with respect to a specified time lag τ_0. A translation of the process by the time lag τ_0 results in a process $z_2(t) \equiv z(t - \tau_0)$ that differs relatively little from the process $z_1(t) \equiv z(t)$. The autocovariance function corresponding to τ_0 is therefore not much smaller than the variance of $z(t)$. The opposite is true for a process whose spectral content is dominated by periods much smaller than τ_0. In particular, for sufficiently large τ, the products $z(t)z(t + \tau)$ are sometimes positive, sometimes negative, and their mean value becomes vanishingly small. Therefore

$$\lim_{\tau \to \infty} R_z(\tau) = 0. \qquad (4.1.27)$$

4.1.6 Cross-Covariance Function and Covariance of Two Stochastic Processes

In some applications (e.g., Section 9.2) it is necessary to describe the relationship between two distinct stochastic processes $z_1(t), z_2(t)$, rather the relationship between two stochastic processes that are identical to within a translation time lag τ, as is the case for Fig. 4.4.

Assume the two processes have zero means. The function

$$R_{z_1 z_2}(\tau) = \lim_{T\to\infty} \frac{1}{T} \int_{-T/2}^{T/2} z_1(t) z_2(t+\tau)\, dt, \tag{4.1.28}$$

is defined as the *cross-covariance function*, or simply the *covariance function*, of the signals. The value of the cross-covariance function for $\tau = 0$, denoted by $\sigma_{z_1 z_2}$, is defined as the *covariance* of the processes $z_1(t),\ z_2(t)$. Recall that the value of the autocovariance of a process $z(t)$ for $\tau = 0$ is the variance σ_z^2. The covariance is the counterpart for a pair of distinct processes $z_1(t), z_2(t)$ of the variance σ_z^2 of a process $z(t)$. The *correlation coefficient* of the two processes $z_1(t), z_2(t)$ is defined as the ratio $\rho_{z_1 z_2} = \sigma_{z_1 z_2}/(\sigma_{z_1}\sigma_{z_2})$. Its largest possible value is unity, and is reached if the two processes are identical. The closer the correlation coefficient is to zero, the less similar are the processes $z_1(t), z_2(t)$. Two processes for which $\rho_{z_1 z_2} = 0$ are said to be *uncorrelated.*

4.2 APPROXIMATE REPRESENTATIONS OF STOCHASTIC PROCESSES

In this section we consider a number of stochastic processes of interest in applications, including Gaussian processes, colored noise, white noise, continuously distributed non-Gaussian processes, and dichotomous noise. Our aim is to discuss their approximate representation as periodic or quasiperiodic sums of harmonic terms with stochastic parameters. These representations are used in Chapter 5 to extend Melnikov theory for the case of stochastic dynamical systems.

4.2.1 Remarks on the Use of Approximate Representations of Processes with Unbounded Marginal Distributions

In Chapter 5 we deal with problems in which a stochastic process must satisfy the technical requirement of boundedness. For this reason, if the process is nominally Gaussian—which, as shown by Eq. A5.14, implies infinitely long tails—we use instead a nearly Gaussian approximating process that has bounded marginal distribution, albeit one with a very long tail. This is justified by our exclusive concern with physically realizable processes. Physical processes for which we use the shorthand description "Gaussian" are in fact nearly Gaussian: they are associated with finite, albeit large numbers of contributions, rather than infinite numbers of contributions, as is assumed in the central limit theorem (Section A5.6). The approximating stochastic processes we present in this section are therefore physically meaningful, as well as having bounded marginal distributions, as is required for the application of the Melnikov approach. Since non-Gaussian processes may be obtained from a Gaussian process by nonlinear transformations (see, e.g., Grigoriu, 1995), these statements also apply to approximations of non-Gaussian processes, as is shown in Section 4.2.4.

The approximate stochastic process representations presented here were originally developed for numerical simulation purposes. However, they have been chosen in this section not on account of their effectiveness for simulation purposes, but rather on account of their mathematical form. This form makes it possible to show that the apparatus of Melnikov theory developed for deterministic systems is also applicable to stochastic dynamical systems (see Chapter 5).

4.2.2 Gaussian Colored Noise. Approximate Representations by Kac-Shinozuka and Bennett-Rice Methods

As indicated earlier, a process is Gaussian if its marginal distribution—its distribution at any specified time t_1—is Gaussian. The term *colored* means that the spectral density of the process varies as a function of frequency. In contrast, noise is termed *white* if the spectral density is the same for all frequencies. In this section we consider a Gaussian process $G(t)$ with unit variance and spectral density $\Psi_0(\omega)$.

4.2.2.1 Approximate Kac-Shinozuka Representation of Gaussian Colored Noise

We show that the process $G(t)$ with zero mean, unit variance, and one-sided spectral density $\Psi_0(\omega)$ can be approximated by the bounded random process with derivatives of all orders

$$G_N(t) = (2/N)^{1/2} \sum_{i=1}^{N} \cos(\omega_i t + \phi_{0i}) \tag{4.2.1}$$

(Kac, 1959, p. 47; Shinozuka, 1971), where the parameter N of the process is finite albeit large, and ϕ_{0i} and ω_i ($i = 1, \dots, N$) are independent, identically distributed random variables with uniform distribution over the interval $[0, 2\pi]$ and probability density function

$$f(\omega) = [1/(2\pi)]\Psi_0(\omega), \tag{4.2.1a}$$

respectively. We refer to the process $G_N(t)$ as *Kac-Shinozuka noise*. Each of its realizations is a quasiperiodic function. The temporal mean and, as will be seen subsequently, the ensemble average of $G_N(t)$ are zero.

The one-sided spectral density of the process $G_N(t)$ is $\Psi_0(\omega)$. We first give a qualitative explanation of this statement. All terms in Eq. 4.2.1 have the same amplitude. Since $f(\omega)$ is proportional to the spectral density $\Psi_0(\omega)$, within frequency intervals where $\Psi_0(\omega)$ is relatively large the number of terms with equal amplitude is, on average, correspondingly large. Therefore, by construction, $G_N(t)$ has a power spectral distribution proportional to $\Psi_0(\omega)$.

We now reproduce the proof that the spectral density is $\Psi_0(\omega)$ (Shinozuka, 1971). Consider the autocovariance function of the stationary process $G_N(t)$:

$$\begin{aligned} R(\tau) &= E[G_N(t)G_N(t+\tau)] \\ &= E[G_N(-\tau/2)G_N(\tau/2)], \end{aligned} \tag{4.2.2}$$

where the translation by $-\tau/2$ is permissible on account of the stationarity of the process. In Eq. 4.2.2 we use the definition of the expectation (the multidimensional generalization of Eq. A5.12a applied to the function of four independent variables $g(\omega_1, \omega_2, \phi, \psi)$):

$$\begin{aligned} E[g(x, y, \phi, \psi)] = \int_0^\infty \int_0^\infty \int_0^{2\pi} \int_0^{2\pi} & g(\omega_1, \omega_2, \phi, \psi) \\ & \times f(\omega_1) f(\omega_2) s(\phi) s(\psi)\, d\omega_1\, d\omega_2\, d\phi\, d\psi, \end{aligned} \tag{4.2.3}$$

where ω_1, ω_2 are random variables on the real line with probability density functions $f(\omega_1), f(\omega_2)$, respectively, and ϕ, ψ are random variables in the interval $(0, 2\pi)$ with uniform probability density functions $s(\phi) = s(\psi) = 1/(2\pi)$. From Eq. 4.2.2,

$$\begin{aligned} R(\tau) &= (2/N) \sum_{i=1}^{N} \sum_{j=1}^{N} E[\cos(-\tfrac{1}{2}\omega_i \tau + \phi_{0i}) \cos(\tfrac{1}{2}\omega_j \tau + \phi_{0j})] \\ &= (1/N) \sum_{i=1}^{N} \sum_{j=1}^{N} E[\cos(\tfrac{1}{2}(\omega_j - \omega_i)\tau + \phi_{0i} + \phi_{0j}) \\ &\qquad + \cos(\tfrac{1}{2}(\omega_i + \omega_j)\tau + \phi_{0j} - \phi_{0i})] \\ &= (1/N) \sum_{i=1}^{N} E[\cos(2\phi_{0i}) + \cos(\omega_i \tau)] \\ &\qquad + \text{ terms of the form } E[\cos(a_{ij}\tau + \phi_{0j} \pm \phi_{0i})] \quad (i \neq j) \\ &= (1/2\pi) \int_0^\infty \cos(\omega\tau) \Psi_0(\omega)\, d\omega. \end{aligned} \tag{4.2.4}$$

To obtain Eq. 4.2.4 we used Eqs. 4.2.3 and 4.2.1a, and the fact that the integrals with respect to ϕ_{0i} of terms of the form $\cos(a_{ij}\tau + \phi_{0i} \pm \phi_{0j})$ $(i \neq j)$ vanish. From Eq. 4.1.21 it follows that the one-sided spectral density of $G_N(t)$ is $\Psi_0(\omega)$. The proof of the statement that the expectation of $G_N(t)$ is zero is similarly based on the use of Eq. 4.2.3.

For large N the process is the sum of a large number of contributions; $G_N(t)$ can therefore be made as close to a Gaussian distribution as desired. That is, given any $G_{\max} > 0$ and $\delta > 0$, there exists N such that $|P_N[G_N(t)] - P[G(t)]| < \delta$ uniformly for all $G(t) < G_{\max}$, where $P_N[G_N(t)]$ is the marginal distribution of $G_N(t)$, and the distribution $P[G(t)] = \lim_{N\to\infty} P_N[G_N(t)]$ is Gaussian. For sufficiently large N the distribution $P_N[G_N(t)]$ is an adequate approximation of $P[G(t)]$, however close the requisite approximation. For practical purposes, for very large N, we may therefore view the nearly Gaussian process $G_N(t)$ as Gaussian.

4.2.2.2 Approximate Bennett-Rice Representation of Gaussian Colored Noise

An alternative approximate representation of a Gaussian process $G(t)$ with zero mean, unit variance, and one-sided spectral density $\Psi_0(\omega)$ is

$$G_N(t) = \sum_{k=1}^{N} a_k \cos(\omega_k t + \phi_{0k}), \tag{4.2.5}$$

where $a_k = [2\Psi_0(\omega_k)\Delta\omega/(2\pi)]^{1/2}$, ϕ_{0k} are uniformly distributed over the interval $[0, 2\pi]$, $\omega_k = k\Delta\omega$, $\Delta\omega = \omega_{\text{cut}}/N$, and ω_{cut} is the frequency beyond which $\Psi_0(\omega)$ vanishes or becomes negligible (the cutoff frequency). The process represented by Eq. 4.2.5 is nearly Gaussian, bounded, and has derivatives of all orders; its one-sided spectral density is $\Psi_0(\omega)$ (Rice, 1954, p. 180), and each of its realizations is periodic.

4.2.3 White Noise. Ornstein-Uhlenbeck Processes. Kac-Shinozuka Approximate Representation of White Noise

We begin this section by defining white noise. We then introduce Ornstein-Uhlenbeck processes, which can be used to represent colored Gaussian noise. Ornstein-Uhlenbeck processes depend upon a parameter c. In the limit $c \to 0$ they approach Gaussian white noise, and can be used for its representation by means of the Kac-Shinozuka, Bennett-Rice, or similar techniqes.

4.2.3.1 Definition of White Noise

White noise, denoted by $\gamma_w G_w(t)$, where γ_w is called the *noise intensity*, is defined by the property that its autocovariance function vanishes for any time lag $\tau > 0$, however small; it may therefore be viewed as an idealization of very "nervously" (as opposed to "sluggishly") fluctuating stochastic processes. The property just stated is written as follows:

$$R(\tau) = \gamma_w^2 \delta(\tau), \tag{4.2.6}$$

where $\delta(\tau)$ is the *Dirac delta function*, defined as

$$\delta(\tau) = 0 \quad \text{for } \tau \neq 0; \; \lim_{\Delta t \to 0} \int_{-\Delta t/2}^{\Delta t/2} \delta(\tau)\, d\tau = 1. \tag{4.2.7}$$

From Eqs. 4.2.7 and 4.1.20 it follows that the spectral density of white noise is

$$\Psi_w(\omega) = \gamma_w^2. \tag{4.2.8}$$

White noise may therefore be alternatively defined by the property that its spectral density is constant, rather than varying as a function of the frequency ω, as is the case for colored noise. It follows from Eqs. 4.1.16 and 4.2.8 that the variance of a white noise process is infinity. This is physically impossible, as is an autocovariance that vanishes for even infinitesimal time lags (Eq. 4.2.6). White noise should therefore be viewed as a mathematical idealization with no counterpart in nature except insofar as it represents approximately stochastic processes with spectra that are nearly constant over long frequency intervals.

4.2.3.2 Ornstein-Uhlenbeck Processes and Their Use for the Approximation of White Noise

Consider the process

$$G(t) = (2c)^{1/2} U(t) \tag{4.2.9}$$

where $U(t)$ is an Ornstein-Uhlenbeck process with autocovariance

$$R_U(s) = (1/2c) \exp(-|s|/c), \tag{4.2.10a}$$

variance $1/2c$ (Eq. 4.1.25), and spectral density

$$\Psi_U(\omega) = a/(1 + c^2\omega^2), \tag{4.2.10b}$$

where $a = 1$ or 2 for the double-sided or one-sided spectrum, respectively (Section 4.1.4.2), and the parameter c is referred to as the correlation time.

Equation 4.2.9 represents a colored process $G(t)$ (i.e., a process with spectral density that varies as a function of frequency) with unit variance. For very small c the spectral ordinate $\Psi_U(\omega)$ is almost constant and very small. The colored noise is approximately equivalent to white noise provided that

$$\gamma_w = (2c)^{1/2}\gamma. \tag{4.2.11}$$

Equation 4.2.11 follows from the condition that, in the limit of vanishing c, the autocovariance of the colored process $\gamma G(t)$ must be the same as the autocovariance of the white noise $\gamma_w G_w(t)$ (Eq. 4.2.6).

The simulation of $\gamma_w G_w(t)$ can be performed by applying Eq. 4.2.1 to the process $\gamma G(t)$ defined by Eq. 4.2.9, Eq. 4.2.10 in which c is small (e.g., $c < 0.1$), and Eq. 4.2.11. Equation 4.2.1 is a sum of a relatively large number of small, independent contributions to the approximating process, and is therefore nearly Gaussian. Comments similar to those made in Section 4.2.1 on the use of the term "Gaussian" for a process that is as nearly Gaussian as desired are also applicable to the term "Gaussian white noise."

In the dimensional counterpart of the stochastic process being considered, if the process represents a force the dimensions of γ_w and $G_w(t)$ are $[FT^{1/2}]$ and $[T^{-1/2}]$, respectively, while the dimension of γ is $[F]$ and $G(t)$ is nondimensional.

4.2.4 Approximate Representation of Non-Gaussian Processes with Continuous Marginal Distributions

Let $X(t)$ denote a non-Gaussian process with zero mean, marginal distribution F, one-sided spectral density $\Psi_X(\omega)$, and variance σ_X^2. Consider a stationary Gaussian process $Y(t)$ with zero mean, unit variance, and spectral density $\Psi_X(\omega)/\sigma_X^2$. Let Φ denote the distribution of the standard Gaussian variable. The process

$$X(t) = F^{-1}[\Phi(Y(t))] \tag{4.2.12}$$

has the marginal distribution F. According to results of numerical simulations, for most processes of interest in applications the spectral density of the function $X(t)$ defined by Eq. 4.2.12 is, to within a close approximation, equal to the original spectrum $\Psi_X(\omega)$ (Grigoriu, 1995).

We represent the process $Y(t)$ by an approximating bounded sum of harmonics $Y_N(t)$. We then approximate Eq. 4.2.12 by a polynomial of degree n. This can be done to any accuracy in a bounded interval. Let

$$X_{n,N}(t) = p_n[Y_N(t)] \tag{4.2.13}$$

be a polynomial approximation of $X(t)$ based on the representation $Y_N(t)$ of $Y(t)$, where

$$p_n(y) = \sum_{i=1}^{n} a_i y^i. \tag{4.2.14}$$

The non-Gaussian noise $X_{n,N}(t)$ follows the marginal distribution F approximately and each of its realizations can be expressed as a linear combination of harmonics. This statement is based on the representation of $Y_N(t)$ as a sum of harmonics, Eqs. 4.2.13 and 4.2.14, and the identity

$$\prod_{i=1}^{n} \cos\beta_i = (1/2)^{n-1} \sum_{p_2,\ldots,p_n=0,1} \cos\left[\beta_1 + \sum_{j=2}^{n}(-1)^{p_j}\beta_j\right] \tag{4.2.15}$$

obtained from the successive application of the identity $\cos\alpha\cos\beta = [\cos(\alpha+\beta)+\cos(\alpha-\beta)]/2$ (Simiu and Grigoriu, 1995).

4.2.5 Representation of Non-Gaussian Processes with Discrete Marginal Probability Distribution

As an example of such processes we consider dichotomous noise, which is characterized primarily by whether the noise is "on" or "off," or "up" or "down." We consider, in particular, coin-toss square-wave dichotomous noise, whose expression is

$$G(t) = a_n \qquad [\alpha + (n-1)]t_1 < t \leq (\alpha + n)t_1 \tag{4.2.16}$$

where n is the set of integers, α is a random variable uniformly distributed between 0 and 1, a_n are independent random variables that take on the values -1 and 1 with probabilities 1/2 and 1/2, respectively, and t_1 is a parameter of the process $G(t)$ (Fig. 4.5).

A rectangular pulse wave of amplitude a_n and length t_1 centered at coordinate $t_n = (\alpha + n - 1/2)t_1$ has Fourier transform $F_n(\omega) = a_n|(2/\omega)\sin(\omega t_1/2)\exp(-j\omega t_n)|$, where $j = \sqrt{-1}$ (see, e.g., Papoulis, 1962, p. 20). The pulse itself can then be expressed as a sum of harmonic terms approximating as closely as desired the inverse Fourier transform of $F_n(\omega)$. Each realization of the coin-toss dichotomous square wave can then be approximated as closely as desired by a superposition of such sums. The Fourier representation just described is admissible provided that the time interval over which the noise realizations are defined is bounded, albeit very large. This restriction, which as remarked in Section 4.1.1 is consistent with the fact that physical processes have finite duration, guarantees the absolute integrability of each realization of the process (Eq. 4.1.3) and, therefore, the existence of its Fourier integral (see Section 4.1.2.2).

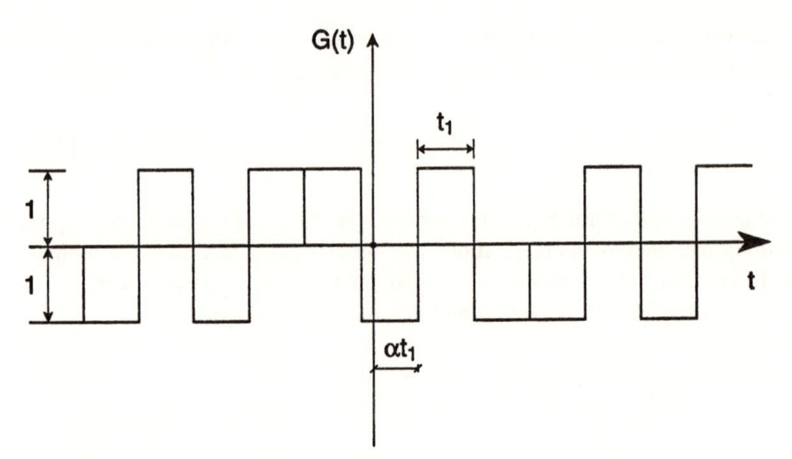

Figure 4.5. Realization of a coin-toss square-wave dichotomous noise.

4.3 SPECTRAL DENSITY OF THE OUTPUT OF A LINEAR FILTER WITH STOCHASTIC INPUT

This section presents material needed in Chapter 5 to show that, for stochastic systems, a simple relation exists between the spectral density of the Melnikov function's stochastic counterpart and the spectral density of the system's excitation process. Section 4.3.1 includes basic general material on linear filters that parallels material applied specifically to the Melnikov function in Sections 2.5.2 to 2.5.4. In Section 4.3.2 we derive the expression for the spectral density of the output of a linear filter whose input is a stochastic process with specified spectral density.

4.3.1 Filters, Impulse Response Function, Convolution Integral, and Transfer Function

The response of a linear system to excitation by the stationary process $P(t)$ may be written as

$$s(t) = \int_{-\infty}^{\infty} h(t-\xi)P(\xi)\,d\xi, \tag{4.3.1}$$

that is, owing to the linearity of the system, the response may be written as a linear superposition of the elemental contributions $h(t-\xi)P(\xi)\,d\xi$. The function $h(t-\xi)$ is the output at time t due to a unit impulse (a delta Dirac function) $\delta(\xi)$ acting at time ξ. We refer to our system as a *noncausal filter* (see Section 2.5.2); to the function $P(t)$ as the *filter input*; and to the response $s(t)$ as the *filter output*. With a change of variable $t' = t - \xi$ we obtain

$$s(t) = \int_{-\infty}^{\infty} h(t')P(t - t')\, dt'. \tag{4.3.2a}$$

The function $h(t)$ is called the filter's *impulse response function.*

The integral of Eq. 4.3.2a is a convolution integral, and $s(t)$ is said to be the result of the convolution of $P(t)$ with $h(t)$. Equation 4.3.2a is also written in the form

$$s(t) = h * P, \tag{4.3.2b}$$

where the symbol $*$ denotes convolution.

Let us consider the particular case of a harmonic input with unit amplitude

$$P(t) = \mathrm{Re}[\exp(j\omega t + \theta)], \tag{4.3.3}$$

where $j = \sqrt{-1}$. Equation 4.3.2 becomes

$$\begin{aligned} s(t) &= \mathrm{Re}[s(t)] \\ &= \mathrm{Re}\left\{ \int_{-\infty}^{\infty} h(t') \exp[j\omega(t - t') + \theta]\, dt' \right\} && (4.3.4a) \\ &= \mathrm{Re}\left[\exp(j\omega t + \theta) \int_{-\infty}^{\infty} h(t') \exp(-j\omega t')\, dt' \right] && (4.3.4b) \\ &= \mathrm{Re}[\boldsymbol{\alpha}(\omega) \exp(j\omega t + \theta)] && (4.3.4c) \\ &= \mathrm{Re}[\boldsymbol{\alpha}(\omega)] \cos(\omega t + \theta) - \mathrm{Im}[\boldsymbol{\alpha}(\omega)] \sin(\omega t + \theta) && (4.3.4d) \\ &= |\boldsymbol{\alpha}(\omega)| \{\cos[\psi(\omega)] \cos(\omega t + \theta) \\ &\quad + \sin[\psi(\omega)] \sin(\omega t + \theta)\} && (4.3.4e) \\ &= |\boldsymbol{\alpha}(\omega)| \cos[\omega t + \theta - \psi(\omega)], && (4.3.4f) \end{aligned}$$

$$\boldsymbol{\alpha}(\omega) = \int_{-\infty}^{\infty} h(t') \exp(-j\omega t')\, dt', \tag{4.3.5a}$$

$$\psi(\omega) = \tan^{-1}\{-\mathrm{Im}[\boldsymbol{\alpha}(\omega)]/\mathrm{Re}[\boldsymbol{\alpha}(\omega)]\}, \tag{4.3.5b}$$

$$\begin{aligned} &\cos[\psi(\omega)] = \mathrm{Re}[\boldsymbol{\alpha}(\omega)]/|\boldsymbol{\alpha}(\omega)|; \\ &\sin[\psi(\omega)] = -\mathrm{Im}[\boldsymbol{\alpha}(\omega)]/|\boldsymbol{\alpha}(\omega)|. && (4.3.5c,d) \end{aligned}$$

The complex function $\boldsymbol{\alpha}(\omega)$ is referred to as the system's *transfer function.* The alternative terms *admittance function* and *receptance* are used in some applications. By Eq. 4.3.5a the transfer function $\boldsymbol{\alpha}(\omega)$ is the Fourier transform of the impulse response function $h(t')$.

4.3.2 Output of Filter with a Stochastic Input: Autocovariance Function and Spectral Density

We now assume that the input $P(t)$ in Eq. 4.3.2a is a stationary stochastic process with zero mean, autocovariance function $R_P(\tau)$, and spectral density $\Psi_P(\omega)$. In this section we obtain expressions for the autocovariance function $R_s(\tau)$ and the spectral density $\Psi_s(\omega)$ of the filter output $s(t)$.

4.3.2.1 Autocovariance Function of $s(t)$

The autocovariance function of the stochastic process with zero mean $s(t)$ is

$$R_s(\tau) = E[s(t)s(t+\tau)] \tag{4.3.6}$$

(Eq. 4.1.24). Using Eq. 4.3.2a we have

$$R_s(\tau) = \int_{-\infty}^{\infty} h(t')\,dt \int_{-\infty}^{\infty} h(t'')E[P(t-t')P(t+\tau-t'')]\,dt'' \tag{4.3.7a}$$

$$= \int_{-\infty}^{\infty} h(t') \int_{-\infty}^{\infty} h(t'')E[P(t)P(t+t'-t''+\tau)]\,dt'\,dt'' \tag{4.3.7b}$$

$$= \int_{-\infty}^{\infty} h(t') \int_{-\infty}^{\infty} h(t'')R_P(t'-t''+\tau)\,dt'\,dt''. \tag{4.3.7c}$$

Equation 4.3.7b follows from the stationarity of the process $P(t)$, by virtue of which a translation of the origin by t' results in no change in the expectation $E[P(t-t')P(t+\tau-t'')]$. We now show that Eq. 4.3.7c can be used to obtain the spectral density of the output process $s(t)$.

4.3.2.2 Spectral Density of $s(t)$

From Eqs. 4.1.20 and 4.3.7c we have

$$\Psi_s(\omega) = \int_{-\infty}^{\infty} R_s(\tau)\exp(-j\omega\tau)\,d\tau \tag{4.3.8a}$$

$$= \int_{-\infty}^{\infty} \left[\int_{-\infty}^{\infty} h(t') \int_{-\infty}^{\infty} h(t'')R_P(t'-t''+\tau)\,dt'\,dt''\right] \times \exp(-j\omega\tau)\,d\tau \tag{4.3.8b}$$

$$= \int_{-\infty}^{\infty} h(t') \int_{-\infty}^{\infty} h(t'')$$

$$\times \left[\int_{-\infty}^{\infty} R_P(t' - t'' + \tau) \exp(-j\omega\tau)\, d\tau \right] dt'\, dt''$$

$$= \int_{-\infty}^{\infty} h(t') \exp(j\omega t')\, dt' \int_{-\infty}^{\infty} h(t'') \exp(-j\omega t'')\, dt'' \qquad (4.3.8c)$$

$$\times \int_{-\infty}^{\infty} R_P(t' - t'' + \tau)$$

$$\times \exp[-j\omega(t' - t'' + \tau)]\, d(t' - t'' + \tau). \qquad (4.3.8d)$$

Equations 4.3.8d, 4.3.5a, and 4.1.20 yield

$$\Psi_s(\omega) = \boldsymbol{\alpha}^*(\omega)\boldsymbol{\alpha}(\omega)\Psi_P(\omega)$$

$$= |\boldsymbol{\alpha}(\omega)|^2 \Psi_P(\omega), \qquad (4.3.9)$$

where $\boldsymbol{\alpha}(\omega)$ is the filter's transfer function, that is, *the spectral density of the output of a linear time-invariant filter with a stationary stochastic process input is equal to the spectral density of the excitation process times the square of the modulus of the filter's transfer function.* Equation 4.3.9 plays a fundamental role in stochastic Melnikov theory—see Chapter 5.

Chapter Five

Chaotic Transitions in Stochastic Dynamical Systems and the Melnikov Process

The purpose of this chapter is to develop the *stochastic Melnikov method*, that is, to extend the Melnikov method for the case of stochastic near-integrable multistable systems.

It was shown in Chapter 3 that deterministic near-integrable planar multistable systems can have irregular steady-state motions sensitive to initial conditions and exhibiting chaotic transitions between preferred regions of phase space (Figs. 3.12, 3.15, and 3.17; see also Fig. 1.1c). Similar motions with chaotic transitions occur in near-integrable multistable planar systems with stochastic excitation (by definition these include systems with a combination of stochastic and deterministic excitations). Experiments on the electronic device known as the radio frequency (rf) Josephson junction reveal strikingly that, qualitatively, periodic excitation and noise can have similar effects on the occurrence of transitions (Kautz, 1985). A comparison between numerically simulated motion exhibiting transitions in a deterministic system (Fig. 5.1a) and its stochastically excited counterpart (Fig. 5.1b) shows that deterministic and stochastic motions are indeed difficult to distinguish visually.

A similar difficulty has been noted, for example, for chemical reaction kinetics. Argoul, Arneodo, and Richetti (1991) noted "the controversy on the deterministic or stochastic character of chemical chaos" in relation to the Belousov-Zhabotinsky (B.-Z.) reaction[1], and argued that "after measuring the largest Lyapounov exponent, whose positivity confirmed sensitive dependence on initial conditions, the demonstration of the determinism was complete, despite objections voiced by certain experts in the kinetics of the B.-Z. reaction."

In fact, the dichotomy between deterministic chaos and stochastic motions implicit in this statement is not warranted. In this chapter we use the Melnikov method to show that, for near-integrable planar multistable systems, the motion can be chaotic not only under deterministic excitation, but

[1] For a simple description of the Belousov-Zhabotinskii (B.-Z.) reaction from a chaotic dynamics point of view see, e.g., Bergé et al. (1984).

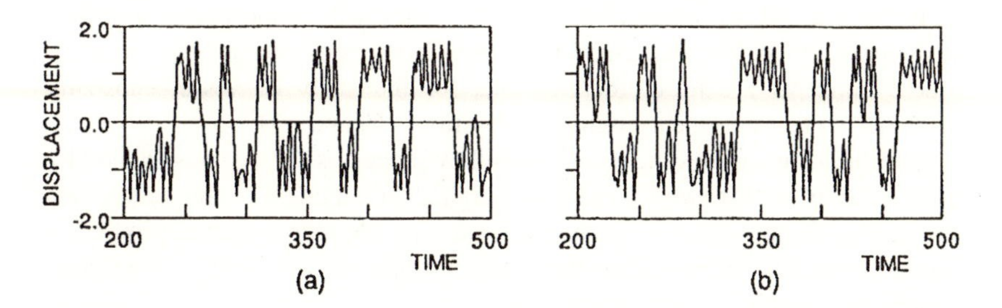

Figure 5.1. Motions with transitions in double-well oscillators. (a) Deterministic, harmonically excited motion. (b) Stochastically excited motion (after Simiu and Frey, 1996a).

under stochastic excitation as well. A positive largest estimated Lyapounov exponent is therefore *not* necessarily indicative of deterministic chaotic motion (Frey and Simiu, 1993; for a particular type of system see also Van den Broeck and Nicolis, 1993).

Essentially, we show that deterministic and stochastic excitations play equivalent roles in the promotion of transitions and chaos. We also show that the stochastic Melnikov method can be used to develop, among other things, criteria that (1) guarantee the nonoccurrence of transitions in systems excited by dichotomous noise or other stochastic processes with bounded marginal distributions, and (2) make it possible to assess the effect on the transition rate of the excitation's spectral density shape.

Our development of the stochastic Melnikov method follows from the fact that the Melnikov necessary condition for chaos may be used for systems with excitation consisting of finite sums of harmonic terms (Sections 2.5.4 and 3.5). To apply the Melnikov condition to stochastic systems we use approximations of the stochastic excitation process by ensembles of realizations (paths), each of which consists of such a sum.

The chapter is organized as follows. In Section 5.1 we show results of a fluidelastic experiment that illustrate the role of stochastic excitation in inducing transitions. Section 5.2 uses approximations presented in Section 4.2 to define and characterize the Melnikov process—the stochastic counterpart of the Melnikov function—for near-integrable systems with Gaussian excitation. In particular, we present the simple but important result that the spectral density of the Melnikov process is equal to the spectral density of the excitation times the square of the Melnikov scale factor. We discuss the application of the Smale-Birkhoff theorem (Section 3.5) to the approximating stochastic systems, and note that, although the stochastically induced transitions are chaotic, episodes may occur during which the chaotic behavior is interrupted; following such episodes the motion settles again on a chaotic course.

In Section 5.3 we extend to stochastic systems the concepts of transport and phase space flux discussed for deterministic systems in Section 2.7.5. These concepts are used, for example, for the study of vessel capsizing (Chapter 6) and nonlinear control applications (Chapter 7).

Section 5.4 examines the use of Melnikov processes for systems excited by stochastic excitations with bounded marginal distributions, such as systems excited by dichotomous noise. In particular, the Melnikov method provides a simple condition guaranteeing the nonoccurrence of transitions in nonlinear systems excited by dichotomous noise.

Section 5.5 uses chaotic transport considerations to show that the mean time between consecutive zero upcrossings of a system's Melnikov process is a lower bound for the system's mean time between consecutive escapes (or, for short, the system's mean escape time). For systems with Gaussian excitation the Melnikov method is used to obtain lower bounds for the probability that transitions will not occur—or upper bounds for the probability that transitions can occur—during specified time intervals.

In Section 5.6 it is shown that the Melnikov scale factor is an indicator of the relative degree to which the various frequency components of the excitation are effective in inducing transitions. This is useful for the qualitative prediction of the effect of the excitation's spectral shape on the transition rate, and has application in control theory and for system modeling and identification.

Section 5.7 is concerned with the application of the stochastic Melnikov method to slowly varying planar systems, and points out limitations of the method that may arise in such applications from the possibility of nonchaotic escapes.

Section 5.8 investigates the escape problem for a special type of system that allows the application—and cross-validation—of two approaches: one that utilizes the Fokker-Planck equation, and one based on the stochastic Melnikov method.

As indicated earlier, our development of the stochastic Melnikov method uses results from the theory of deterministic chaos in conjunction with approximate representations of stochastic processes. An alternative approach based on the construction of stochastic generalizations of symbolic dynamics techniques is discussed by Gundlach (1995) and references quoted therein; see also Steinkamp (1999).

5.1 BEHAVIOR OF A FLUIDELASTIC OSCILLATOR WITH ESCAPES: EXPERIMENTAL AND NUMERICAL RESULTS

The fluidelastic experiment presented in this section illustrates the role played by stochastic excitation in inducing transitions in multistable dynamical sys-

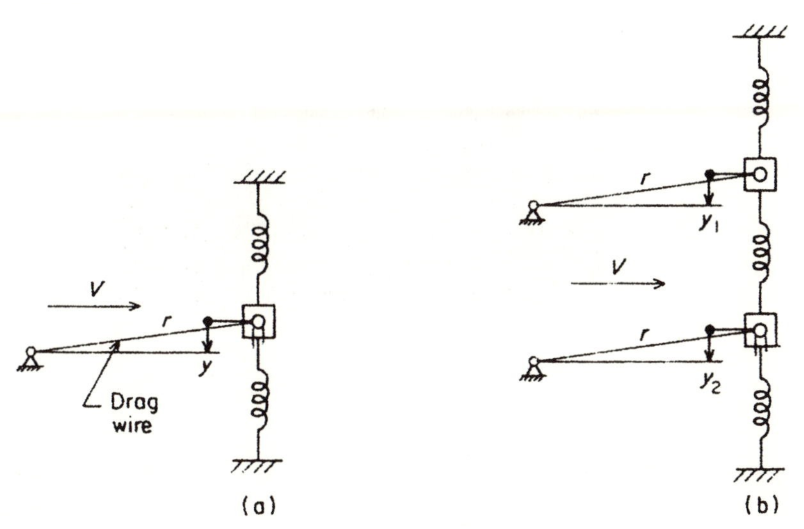

Figure 5.2. Schematics of (a) one-prism galloping oscillator, (b) two-prism galloping oscillator.

tems. A fluidelastic oscillator is an elastically restrained oscillator immersed in a flow with which it interacts nonlinearly. The experimental device we consider for our illustration is a double-prism galloping oscillator.

Before describing the device we consider the single-prism galloping oscillator, that is, a horizontal prism with square cross section restrained elastically in a vertical plane. The prism is immersed in a smooth horizontal flow with velocity normal to its longitudinal axis (Fig. 5.2a). If the prism is at rest, the relative velocity of the flow with respect to the prism is horizontal and, on account of symmetry, the prism experiences no steady lift force. However, if the prism has a nonzero vertical velocity, the relative velocity of the flow with respect to the prism is no longer horizontal, and the prism experiences a lift force. In the absence of other forces, this lift force would induce periodic galloping, that is, periodic oscillations occurring approximately in a vertical plane (Parkinson and Smith, 1964; Simiu and Scanlan, 1996, p. 230).

In reality the prism is subjected to additional lift forces associated with vorticity shed in its wake and with secondary flows due to end plates and other experimental appurtenances. For this reason the oscillations are irregular, rather than periodic. This is confirmed by results of water tunnel experiments and numerical simulations. A record of one-prism oscillator motion observed in a water tunnel is shown in Fig. 5.3 (Simiu and Cook, 1992).

We now consider the double-prism galloping oscillator (Fig. 5.2b). Experimental observations of two-prism oscillator motions show that, for relatively low flow velocities, the two prisms gallop in phase. For higher velocities,

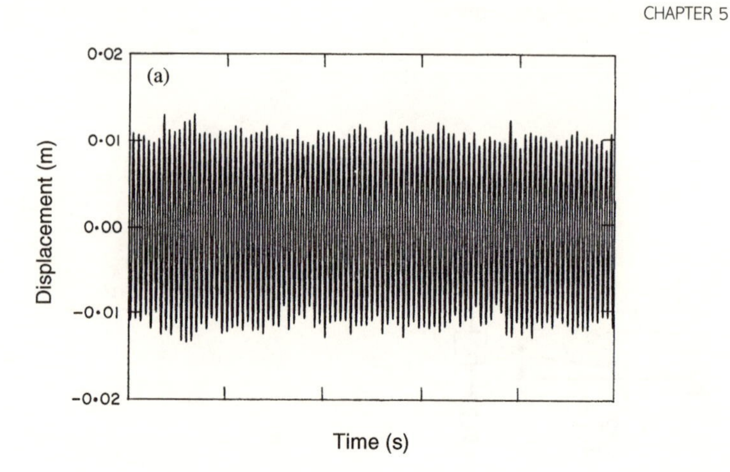

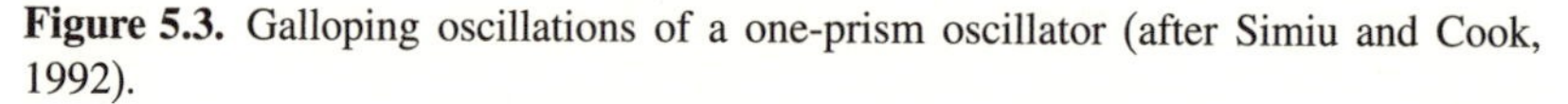

Figure 5.3. Galloping oscillations of a one-prism oscillator (after Simiu and Cook, 1992).

however, there occur irregular transitions from this galloping pattern to one wherein the prisms oscillate in opposite phases, and conversely. The transition rate increases with increasing flow velocity. Records of observed galloping motions with transitions are shown in Fig. 5.4a for the top prism and in Fig. 5.4b for the top and bottom prisms.

Motions of numerically simulated stochastically excited models of the two-prism galloping oscillator were found to be similar to the observed motions of Fig. 5.4 (Simiu and Cook, 1992). On the other hand, in the absence of stochastic lift forces, simulations were unable to reproduce the observed behavior. This suggests that the observed transitions are induced by the type of stochastic forces to which we ascribe the irregularities seen in Fig. 5.3.

5.2 SYSTEMS WITH ADDITIVE AND MULTIPLICATIVE GAUSSIAN NOISE: MELNIKOV PROCESSES AND CHAOTIC BEHAVIOR

A system excited by a stationary stochastic process may be viewed as an ensemble of systems, each of which is excited by a different realization of that process. The ensemble of realizations of the stochastic process can be approximated by an ensemble of sums of finite numbers of harmonic terms with random parameters (Section 4.2). Each member of the ensemble of systems then possesses a Melnikov function. The ensemble of Melnikov functions induced by the approximating stochastic process is a Melnikov process.

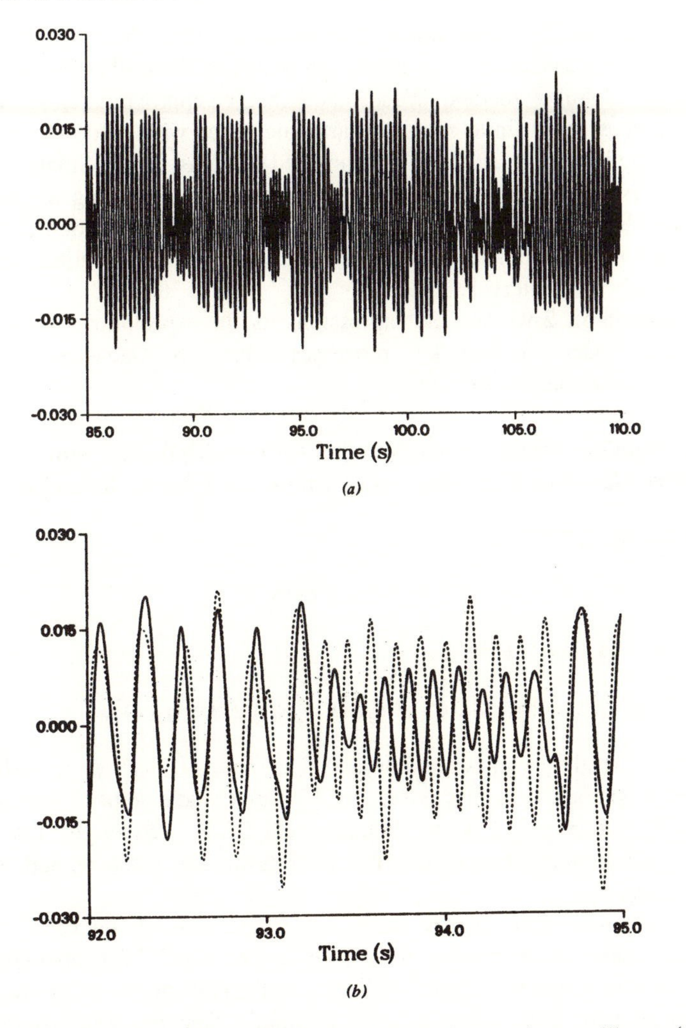

Figure 5.4. Observed oscillations with transitions of two-prism oscillator: (a) top prism; (b) top prism (solid line) and bottom prism (dotted line) (after Simiu and Cook, 1992).

In Section 5.2.1 we obtain expressions for Melnikov processes for systems excited by *additive* Gaussian noise (that is, a Gaussian process independent of the system's state) or *multiplicative* Gaussian noise (that is, the product of a state-dependent function and a Gaussian process independent of the system's state). Like the expressions for the Melnikov functions induced in deterministic systems by additive and multiplicative excitation, the expressions for the Melnikov processes are essentially the same for systems with additive and multiplicative noise. The simplicity with which it handles systems with multiplicative noise excitation is one of the useful features of the stochastic Melnikov method. We also note that the approximation of the excitation processes by finite sums of harmonic terms with random parameters allows application to stochastically excited systems of the Smale-Birkhoff theorem (see Section 3.5). Motions with transitions occurring in such systems are therefore chaotic and, hence, sensitive to initial conditions, although interruptions of the chaotic motion by large excursions of the noise can occur at relatively long intervals.

In Section 5.2.2 we provide expressions for the expectation, spectral density, and variance of Melnikov processes induced by Gaussian excitations with specified spectral density.

5.2.1 Melnikov Processes for Systems with Multiplicative and Additive Gaussian Excitation. Transitions and Chaotic Behavior

Consider the system

$$\dot{\mathbf{x}} = f(\mathbf{x}) + \epsilon g(\mathbf{x}, t), \tag{5.2.1}$$

where $\mathbf{x} = (x_1, x_2)^{\mathrm{T}}$, $\mathbf{f} = (f_1, f_2)^{\mathrm{T}}$,

$$\mathbf{g}(\mathbf{x}, t) = \{\gamma_1(\mathbf{x})G(t) + q_1(\mathbf{x}), \gamma_2(\mathbf{x})G(t) + q_2(\mathbf{x})\}, \tag{5.2.2}$$

the unperturbed system is Hamiltonian (i.e., $f_1 = \partial H/\partial x_2$, $f_2 = -\partial H/\partial x_1$) and has a homoclinic orbit $\mathbf{x} = [x_{h1}, x_{h2}]^T$ with a saddle point O at the origin, $0 < \epsilon \ll 1$, $\mathbf{f}$, $\gamma_k(\mathbf{x})$, $q_k(\mathbf{x})$ are bounded and C^r, $r \geq 2$ (see Section 2.4.1), and $G(t)$ is a stochastic process with zero mean, unit variance, and spectral density $\Psi_0(\omega)$.

Provided that $G(t)$—that is, each of its realizations—is bounded, each of the realizations of the process $\mathbf{g}(\mathbf{x}, t)$ defined by Eq. 5.2.2 is also bounded. Then, as was shown in Section 2.4.1, each of the realizations of the stochastic dynamical system 5.2.1 possesses a Melnikov function. The ensemble of these Melnikov functions is the Melnikov process induced in the dynamical system by the stochastic excitation process.

In general stochastic processes $G(t)$ are not bounded. In particular, this is true of Gaussian stochastic processes, whose marginal distributions have infinitely long tails, and of non-Gaussian processes with continuous, infinitely tailed marginal distributions, which can be obtained from a Gaussian distribution by nonlinear transformations. However, as was indicated in Section 4.2.1, Gaussian processes are an idealized representation of processes that are in fact bounded. The idealization implies that the process is due to an infinitely large number of contributions. In physically realizable systems that number is finite, albeit very large, and it is common in engineering, physics, and the life sciences to approximate—or simulate—unbounded stochastic processes by bounded processes with continuous derivatives of all orders (see Section 4.2).

In the particular case of Gaussian processes $G(t)$ we may use the approximation $G(t) \approx G_N(t)$, where

$$G_N(t) = (2/N)^{1/2} \sum_{i=1}^{N} \cos(\omega_i t + \phi_{0i})$$

(Eq. 4.2.1) where ϕ_{0i} and $\omega_i (i = 1, \ldots, N)$ are independent, identically distributed random variables with uniform distribution over the interval $[0, 2\pi]$ and probability density function $f(\omega) = [1/(2\pi)]\Psi_0(\omega)$, respectively, and $\Psi_0(\omega)$ is the spectral density of $G(t)$. As was shown in Section 4.2.1, the spectral density of the process $G_N(t)$ is $\Psi_0(\omega)$. Results similar to those obtained by using the approximation 4.2.1 would be obtained by using other approximations consisting of sums of harmonic functions with random parameters (e.g., Eq. 4.2.5 or, if the noise is white, the approximate noise representation discussed in Section 4.2.3).

It follows from the results of Section 2.4.1 that the process $G_N(t)$ induces a Melnikov process

$$M(t_0) = \int_{-\infty}^{\infty} h(\zeta) G_N(t_0 - \zeta)\, d\zeta - k \tag{5.2.3}$$

$$= h^* G_N - k, \tag{5.2.4}$$

where $h(\zeta)$ denotes the Melnikov filter's impulse response function (Eq. 2.5.5d), k is a constant defined by Eq. 2.5.3, and h^*G_N denotes the convolution of h and G_N. Since the Melnikov process is a linear transformation of a near-Gaussian excitation, it is itself near-Gaussian.

A realization of the process $G_N(t)$ corresponds to a fixed set of values of the random parameters that characterize the approximating stochastic process. It follows from Eqs. 2.5.22 and 2.5.23 (Section 2.5.4) that each realization of the time-dependent part of the Melnikov process $M(t_0)$ is a sum of harmonic functions. The entire apparatus of the Melnikov method, including the Smale-Birkhoff theorem, is therefore applicable to the stochastic dynamical systems defined by Eq. 5.2.1. This is the case regardless of whether the excitation is multiplicative or additive since, as was shown in Section 2.5.2, the difference between these two cases is absorbed in the expression for the Melnikov impulse response function $h(t)$.

For any realization of a Gaussian excitation approximated by $G_N(t)$ the necessary condition for transitions and chaotic behavior—that the associated realization of the Melnikov process have simple zeros—is always satisfied for sufficiently large N and over a sufficiently long time interval. Therefore, no matter how small the parameter ϵ in Eqs. 5.2.1, escapes from a potential well can occur for any realization of the Gaussian excitation, although the time between escapes can be long. This result has long been known. The result that motion in near-integrable systems with transitions induced by Gaussian excitation is chaotic—implying sensitivity to initial conditions—is relatively recent (Simiu, Frey, and Grigoriu, 1991; Frey and Simiu, 1993). We note, however, that, as will be shown in Section 5.5.3.1, the chaotic motion may experience interruptions due to large excursions of the noise.

5.2.2 Expectation, Spectral Density, and Variance of Melnikov Processes for Systems with Additive or Multiplicative Excitation

From Eqs. 5.2.3 and 4.2.1 it follows that the Melnikov process of a system with Gaussian excitation has expectation

$$E[M(t_0)] = -k. \tag{5.2.5}$$

For the system 5.2.1 in which $G(t)$ is a Gaussian process with unit variance and spectral density $\Psi_0(\omega)$, it is easy to see from Eq. 4.3.9 that the spectral density of the Melnikov function is

$$\Psi_M(\omega) = |\boldsymbol{\alpha}(\omega)|^2 \Psi_0(\omega). \tag{5.2.6}$$

Remark. From Eqs. 5.2.2, 2.5.5d, and 4.3.5a it can be seen that the functions $\gamma_1(\mathbf{x})$, $\gamma_2(\mathbf{x})$ are absorbed in the expression of the Melnikov scale factor $|\boldsymbol{\alpha}(\omega)|$. If $\gamma_1(\mathbf{x}) = 0$, $\gamma_2(\mathbf{x}) = \gamma =$ const and the variance of $G(t)$ is unity, γ represents the standard deviation of the excitation $\gamma G(t)$. The Melnikov scale factor is then defined as the function $|\boldsymbol{\alpha}(\omega)|$ corresponding to $\gamma = 1$, and in Eq. 5.2.6 the spectrum $\gamma^2 \Psi_0(\omega)$ is used instead of $\Psi_0(\omega)$ (see Remark at the end of Section 2.5.3).

If we choose to consider the one-sided spectral density of $G(t)$, the variance of the Melnikov process is

$$\sigma_M^2 = \frac{1}{2\pi} \int_0^\infty \Psi_M(\omega)\, d\omega. \tag{5.2.7}$$

If we considered instead the two-sided spectral density function of $G(t)$, the lower integration limit in Eq. 5.2.7 would be $-\infty$ instead of 0.

5.3 PHASE SPACE FLUX

It was seen in Section 2.7 that, for a class of near-integrable deterministic planar systems excited by a sum of harmonic terms, escapes from a potential

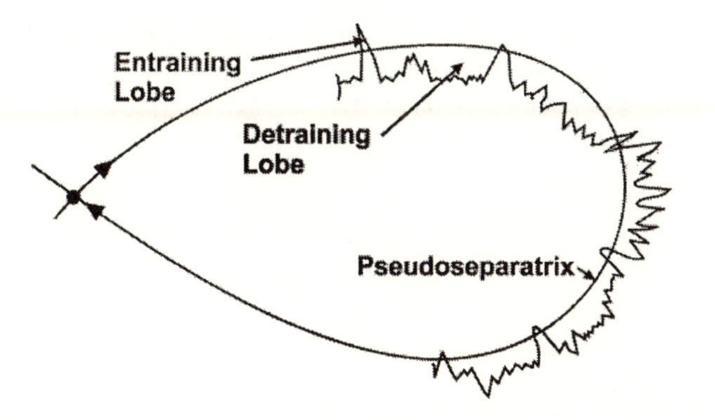

Figure 5.5. Schematic of phase space slice through intersecting manifolds of stochastic system.

well are associated with the mechanism of transport across the pseudoseparatrix. The same mechanism is at work for those systems' stochastically excited counterparts, since each of the realizations of the stochastic excitation can be approximated by a sum of harmonic terms. Figure 5.5 shows, schematically, a phase space slice through the stable and unstable manifolds induced in a system by a realization of its stochastic excitation. As is the case for the deterministic periodic or quasiperiodic excitation case (see, e.g., Figs. 2.10 and 2.11), the phase space slice has a pseudoseparatrix, entraining and detraining lobes, and turnstile lobes.

The phase space flux factor Φ was defined for deterministic systems by Eq. 2.7.4. The same definition applies to those systems' stochastic counterparts, except that $M(t_0)$ denotes the Melnikov process induced by the stochastic excitation, rather than the deterministic system's Melnikov function. For the system

$$\dot{x}_1 = x_2, \tag{5.3.1a}$$

$$\dot{x}_2 = -V'(x_1) + \epsilon[-\beta x_2 + \gamma G(t)], \quad \beta > 0, \quad 0 \leq \epsilon \ll 1, \tag{5.3.1b}$$

where $G(t)$ is a Gaussian process with unit variance, the expression for the flux factor is

$$\Phi = \sigma_M f(k/\sigma_M) - k[1 - F(k/\sigma_M)]. \tag{5.3.2}$$

In Eq. 5.3.2 f and F denote the standard normal probability density function and cumulative distribution function. We remark that Φ is a constant independent of the realization of the stochastic process being considered.

Equation 5.3.2 can be proved by invoking the ergodicity of the process $[M(t)]^+$. We equate the temporal mean (Eq. 2.7.4) with the ensemble average, and use the notation $M(t) = M_0(t) - k$. (The process $M_0(t)/\sigma_M$ is Gaussian with zero mean and unit standard deviation.) We have

$$\begin{aligned}\Phi = E[M^+(\xi)] &= \sigma_M E[(M_0(\xi)/\sigma_M - k/\sigma_M)^+] \\ &= \sigma_M \int_{k/\sigma_M}^{\infty} (z - k/\sigma_M) f(z)\, dz,\end{aligned} \tag{5.3.3}$$

from which Eq. 5.3.2 follows if it is noted that

$$z f(z)\, dz = -1/(2\pi)^{1/2} \exp(-z^2/2) d(-z^2/2). \tag{5.3.4}$$

For a system excited by a Gaussian process with fixed spectral shape, the escape rate $1/\tau_\epsilon$ increases as the noise intensity γ increases. Figure 5.6a shows that the phase space flux factor Φ also increases as γ increases (recall that σ_M is proportional to γ, as follows from Eqs. 5.2.6, 5.2.7, and 2.5.5d). It may therefore be expected that the flux factor can serve as a measure of the escape rate. This is suggested by Fig. 5.6b, which shows the relation between Φ and τ_ϵ obtained by Monte Carlo simulation for a standard Duffing–Holmes

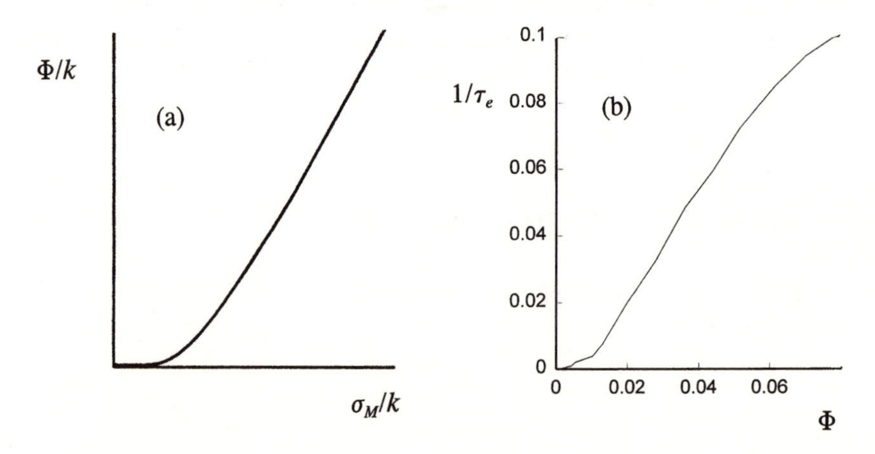

Figure 5.6. (a) Phase space flux factor Φ for system 5.3.1 with Gaussian excitation; k and σ_M are the mean and the standard deviation of the Melnikov process. (b) Example of relation between flux factor and escape rate $1/\tau_\epsilon$.

oscillator and a spectral density $\Psi_0(\omega) = 2\pi/5$ in the interval $0 \le \omega < 5$ and $\Psi_0(\omega) = 0$ elsewhere.

5.4 CONDITION GUARANTEEING NONOCCURRENCE OF ESCAPES IN SYSTEMS EXCITED BY FINITE-TAILED STOCHASTIC PROCESSES. EXAMPLE: DICHOTOMOUS NOISE

Melnikov processes induced by finite-tailed stochastic excitations (i.e., by stochastic processes whose marginal distributions have finite tails) are also finite tailed. Therefore, parameter regions exist for which the Melnikov necessary condition for chaos cannot be satisfied. A criterion guaranteeing the nonoccurrence of chaos and, therefore, of chaotic transitions from a potential well, follows from the requirement that the system parameters be contained in those regions. As an example, we consider the standard Duffing–Holmes oscillator, that is, Eqs. 5.3.1 in which $V(x_1)$ is defined by Eq. 2.1.7, and $G(t)$ is the coin-toss square-wave dichotomous noise process defined by

$$G(t) = a_n, \quad [\alpha + (n-1)]t_1 < t \le (\alpha + n)t_1, \tag{5.4.1}$$

n is the set of integers, α is a random variable uniformly distributed between 0 and 1, a_n are independent random variables that take on the values -1 and 1 with probabilities $\frac{1}{2}$ and $\frac{1}{2}$, respectively, and t_1 is a parameter of the process $G(t)$—see Fig. 4.5.

Let $G_a(t)$ be a close approximation of $G(t)$ that satisfies the conditions required for the proof of the persistence theorem (Section 2.3.2). The Melnikov process induced in the system by $G_a(t)$ is

$$M_a(t_0) = \int_{-\infty}^{\infty} h(\zeta) G_a(t_0 - \zeta)\, d\xi - k \tag{5.4.2a}$$

$$= \int_{-\infty}^{\infty} h(\zeta)\{G(t_0 - \zeta) - [G(t_0 - \zeta) - G_a(t_0 - \zeta)]\}\, d\zeta - k. \tag{5.4.2b}$$

where $h\ (t)$ is defined by Eq. 2.5.5d. Since the contribution of the difference $G(t_0 - \zeta) - G_a(t_0 - \zeta)$ to the integral of Eq. 5.4.2b can be made as small as desired, we can write

$$M_a(t_0) \approx M(t_0) \tag{5.4.3}$$

where, formally,

$$M(t_0) = \int_{-\infty}^{\infty} h(\zeta)G(t_0 - \zeta)d\zeta - k. \tag{5.4.4}$$

Therefore, when writing the Melnikov necessary condition for chaos, we may use Eq. 5.4.4; there is no need to develop an explicit expression for $G_a(t)$ and calculate an approximation $M_a(t_0)$.

For the standard Duffing–Holmes oscillator $h(t) = \gamma x_{h2}(-t) = \sqrt{2}\gamma\times$ $\operatorname{sech} t \tanh t$ (Eqs. 2.5.5d, 5.3.1, and 2.5.16 in which $a = b = 1$), and $k = 4\beta/3$ (Eq. 2.5.18). Then, from Eq. 5.4.4,

$$M(t) = -4\beta/3 + \sqrt{2}\gamma F(t), \tag{5.4.5}$$

$$F(t) \approx \sum_{n=-\ell}^{\ell} a_n\{-\operatorname{sech}[(n+\alpha)t_1 - t] + \operatorname{sech}[(n+\alpha-1)t_1 - t]\}, \tag{5.4.6}$$

where ℓ is sufficiently large for the error due to the assumption that ℓ is finite to be as small as desired. The necessary condition for chaos—and escapes—is that $M(t)$ have simple zeros; therefore chaos cannot occur if in Eq. 5.4.5 $M(t) < 0$, that is, if

$$F(t) < 4\beta/(3\sqrt{2}\gamma) = 0.9428\beta/\gamma. \tag{5.4.7}$$

The area under the curve $\operatorname{sech} t \tanh t$ in a half-plane is $\operatorname{sech} t|_0^\infty = 1$. Since the sum in Eq. 5.4.6 yields $F(t) = 2$ if $\alpha = 0$, $a_n = 1$ for all n such that $t > 0$, and $a_n = -1$ for all n such that $t < 0$, it follows that $-2 < F(t) < 2$, or $|F(t)| < 2$. Therefore chaos cannot occur if

$$\gamma/\beta < 0.4714. \tag{5.4.8}$$

The simplicity of Eq. 5.4.8 is noteworthy.

Time histories of the function $F(t)$ for $t_1 = 1.0, 0.35$, and 0.1 are shown in Fig. 5.7. It is seen that, as a criterion guaranteeing the nonoccurrence of exits, Eq. 5.4.8—based as it is on the inequality $|F(t)| < 2$—is increasingly weak as t_1 becomes smaller. We remark that Eq. 5.4.8 can also be applied for coin-toss dichotomous noise with random arrival times. More generally, criteria similar to Eq. 5.4.8 can be derived for any reasonable models of tail-limited random excitation.

We show in Figs. 5.8a and 5.8b time history realizations corresponding to the dichotomous noise 5.4.1 with parameters $t_1 = 1.0$, $\epsilon = 0.1$, $\beta = 1.5$, and, respectively, $\gamma/\beta = 0.469 < 0.4714$, $\gamma/\beta = 1.887$. The motion of Fig. 5.8a is confined to one well. Its irregularity is due to the

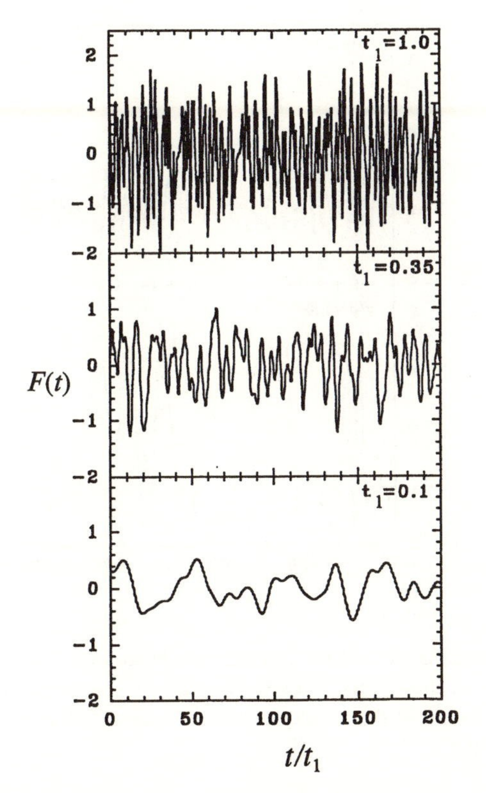

Figure 5.7. Realizations of function $F(t)$ for $t_1 = 1.0$, 0.35, and 0.1. Note that the peaks of $F(t)$ decrease for smaller t_1 (after Sivathanu, Hagwood, and Simiu, 1995).

stochastic nature of the excitation. The chaotic motion of Fig. 5.8b is similar to chaotic motions induced in the standard Duffing–Holmes oscillator by harmonic or quasiperiodic excitation. Its irregularity is due to both the chaotic nature of the motion and the stochastic nature of the excitation. Figure 5.8.b shows that, as for systems with harmonic forcing, for systems with dichotomous noise the Melnikov necessary condition for chaos is helpful in the search for chaotic regions of parameter space even for relatively large perturbations. Sensitivity to initial conditions (i.e., the positivity of the largest Lyapounov exponent) was verified numerically for the motion of Fig. 5.8b.

We have so far assumed that the dichotomous noise excitation is additive. Excitation by multiplicative noise $\gamma(x_1, x_2)G(t)$ merely entails replacing the constant γ by the function $\gamma(x_1, x_2)$ in the expression for the impulse response function $h(t)$ (see Eq. 2.5.5d).

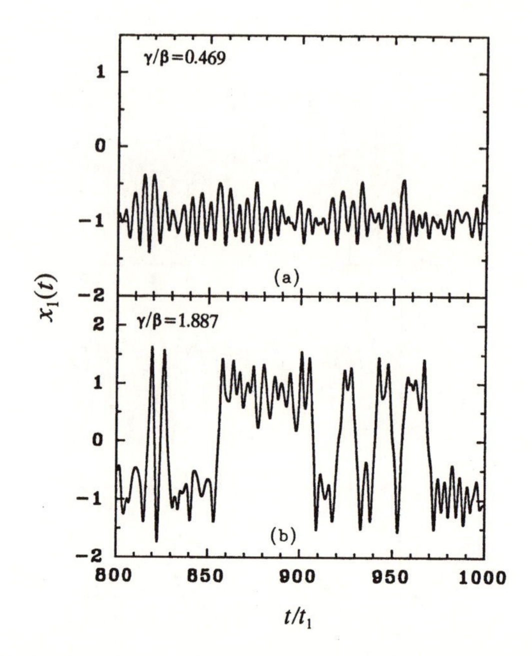

Figure 5.8. (a) Nonchaotic and (b) chaotic time histories $x_1(t)$. The irregularity of (a) is due to the stochastic nature of the excitation. The irregularity of (b) is due to the stochastic nature of the excitation *and* the chaotic nature of the motion (after Sivathanu et al. 1995).

5.5 MELNIKOV-BASED LOWER BOUNDS FOR MEAN ESCAPE TIME AND FOR PROBABILITY OF NONOCCURRENCE OF ESCAPES DURING A SPECIFIED TIME INTERVAL

Escapes from preferred regions associated with the potential wells[2] are due to chaotic transport across the pseudoseparatrix. The transport is effected by the turnstile lobes of the system's intersecting stable and unstable manifolds. The transport mechanism was illustrated in Fig. 2.10 for harmonic excitation, and is similar for stochastically excited systems (see Fig. 5.5). From Fig. 5.5 it follows that on average no transport across the pseudoseparatrix can occur during a time interval smaller than the mean time between consecutive zero upcrossings of the Melnikov process, denoted τ_u. Therefore, the mean time τ_u is a lower bound for the system's mean time of escape from a well, τ_ϵ.

[2] Or, for short, escapes from potential wells. Implicit in this latter term is an approximation, insofar as the pseudoseparatrix of the perturbed system and the separatrix of the unperturbed system do not coincide.

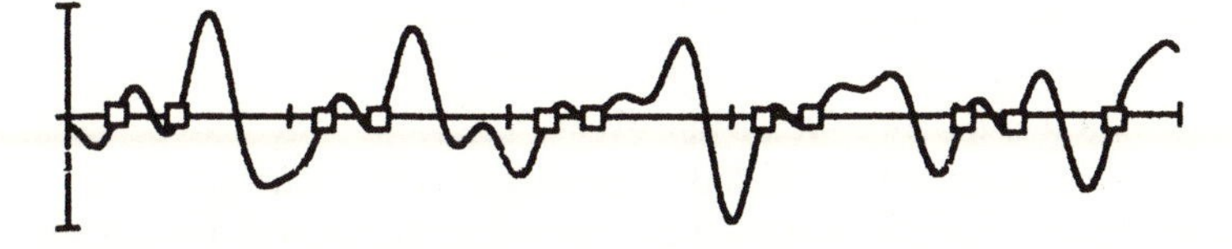

Figure 5.9. Consecutive zero upcrossings (indicated by square symbols) for a realization of a stochastic process with zero mean.

(A zero upcrossing is the intersection of a path $x(t)$ with the axis $x = 0$ such that, at the point of intersection, $dx/dt > 0$—see Fig. 5.9.)

In Section 5.5.1, for systems with Gaussian excitation, we use classical results of the theory of stochastic processes to calculate τ_u and obtain Melnikov-based lower bounds for the probability that no escapes will occur during a specified time interval (upper bounds for the probability that escapes can occur). In Section 5.5.2 we present estimates of the system's mean escape time τ_ϵ for the case of white noise excitation. In Section 5.5.3 we present examples of calculations of mean times τ_u and τ_ϵ for systems excited by white noise, colored Gaussian noise, and dichotomous noise, and of lower bounds for the probability that no escape will occur during a specified time interval in systems with Gaussian excitation. We also comment on the existence of nonchaotic episodes in systems with chaotic motion induced by Gaussian excitation.

5.5.1 Systems with Gaussian Noise: Mean Time between Consecutive Zero Upcrossings of Melnikov Process. Lower Bound for Probability of No Escape during Specified Time Interval

For the system

$$\ddot{x} = -V'(x) + \epsilon(\gamma G(t) - \beta\dot{x}) \qquad 0 < \epsilon \ll 1,\ \beta > 0 \qquad (5.5.1)$$

(which may also be written in the form 5.3.1), the Melnikov process induced by Gaussian noise is Gaussian with mean, spectral density, and variance $-k$, $\Psi_M(\omega)$, and σ_M^2, respectively (Eqs. 5.2.5, 5.2.6, 5.2.7). The mean time between consecutive zero upcrossings for the Melnikov process is

$$\tau_u = \nu^{-1}\exp(\kappa^2/2), \qquad (5.5.2)$$

$$\nu = (1/2\pi)\left\{\left[\int_0^\infty \omega^2\Psi_M(\omega)\,d\omega\right]\Big/\left[\int_0^\infty \Psi_M(\omega)\,d\omega\right]\right\}^{1/2},$$

$$\kappa = k/\sigma_M \qquad (5.5.3\text{a,b})$$

(see Appendix A6). As Eqs. 5.5.3 show, κ is the inverse of the coefficient of variation of the Melnikov process, and ν is proportional to the radius of gyration of the spectrum of the Melnikov process with respect to the origin of the frequency axis.

Provided that the escapes are rare events, the probability that there will be no zero upcrossings of the Melnikov process during a time interval $T \ll \tau_u$ can be closely approximated by the Poisson distribution with average rate of arrival τ_u^{-1} and a number of upcrossings during the interval T equal to zero, that is,

$$P_M(0, T) = \exp(-T/\tau_u) \tag{5.5.4}$$

(Eq. A5.13). The probability $P_M(0, T)$ is a lower bound for the probability that no escapes from a well will occur during the interval T. The probability that there will be at least one zero upcrossing of the Melnikov process during the interval T is

$$p_{M,T} = 1 - \exp(T/\tau_u) \tag{5.5.5}$$

The probability $p_{M,T}$ is an approximate upper bound for the probability that escapes from a well can occur during the interval T.

5.5.2 Mean Escape Time for Planar Multistable Systems Excited by White Noise

If $G(t)$ is a zero mean white noise process with autocovariance function $\delta(t)$ (i.e., with spectral density function equal to unity for all frequencies—see Eq. 4.1.20 in which $R_z(\tau) = \delta(\tau)$), the spectral density of the excitation $\epsilon\gamma G(t)$ is $\Psi_0 = (\epsilon\gamma)^2$. To calculate the mean escape time τ_ϵ we use the following expression for the stationary joint distribution of the variables $x, \dot{x}$:

$$f_{x\dot{x}}(x, \dot{x}) = \text{const} \times \exp(-2\epsilon\beta H/\Psi_0) \tag{5.5.6}$$

(Soong and Grigoriu, 1993, p. 219), where $H(x, \dot{x}) = V(x) + \dot{x}^2/2$ is the system's total energy. Equation 5.5.6 yields

$$\tau_\epsilon = [\epsilon/(4\pi\beta)]^{-1/2}(\gamma a)^{-1} \exp\{[2\beta/(\epsilon\gamma^2)]V(0)\}, \tag{5.5.7a}$$

$$a = \left\{ \int_0^\infty \exp\{-[2\beta/(\epsilon\gamma^2)]V(x)\}\, dx \right\}^{-1}. \tag{5.5.7b}$$

For details on Eqs. 5.5.6 and 5.5.7 see Appendix A7.

5.5.3 Numerical Examples

The examples that follow pertain to the standard Duffing–Holmes oscillator (Eqs. 5.3.1, 2.1.7). In Sections 5.5.3.1 and 5.5.3.2 we consider Gaussian and dichotomous excitation, respectively. We recall that the Melnikov scale factor is defined by Eq. 2.5.17 in which $\gamma = 1$ (see Remarks at the end of Section 2.5.3 and following Eq. 5.2.6).

5.5.3.1 System with Gaussian Excitation

The results of Fig. 5.10 were obtained for white noise and colored noise (4.2.9) with one-sided spectral density (4.2.10b). The smaller the parameter c in Eq. 4.2.10b the closer the noise is to white noise—see Fig. 5.11. Figures 5.10a, 5.10b, and 5.10c correspond, respectively, to white noise or colored noise with $c = 0.1$, colored noise with $c = 0.5$, and colored noise with $c = 2.5$. In all cases $\epsilon = 1$. The mean zero upcrossing rates $1/\tau_u$ are based on Eq. 5.5.2. They depend solely on the ratio β/γ, and are shown in solid lines. The mean escape rates $1/\tau_\epsilon$ are based on Eqs. 5.5.7 for white noise and on Monte Carlo simulations for colored noise. Note in Fig. 5.10a that they are almost indistinguishable for white noise (dashed lines) and colored noise with $c = 0.1$. For fixed β/γ, $1/\tau_\epsilon$ increases as β increases (i.e., as the strength of the perturbation increases).

As β/γ increases, the discrepancy between τ_u and τ_ϵ increases exponentially, and τ_u becomes less and less useful as a lower bound for τ_ϵ. Usually, the lower bound τ_u is therefore useful only for relatively small ratios β/γ. It should be remembered, however, that for very small β/γ (very large γ) the lower bound τ_u breaks down. This is because the excitation is so strong that the system no longer behaves chaotically. (A similar situation occurs if the excitation is harmonic. In that case for small amplitudes of the excitation the motion is periodic and confined to a well. Following a number of bifurcations that occur as the amplitude of the harmonic excitation increases, a set of intermediate amplitudes exists for which the motion is chaotic with transitions between wells. As the amplitude increases still further the motion is no longer chaotic; rather, it visits both wells periodically—see Fig. 1.2.) In fact, although the motion is indeed chaotic as long as γ is not too large, whenever relatively rare short and large outbursts of noise occur the steady-state chaotic motion is interrupted. Following such an interruption the motion again settles on a steady-state chaotic course—until another interruption of the chaotic motion occurs.

5.5.3.2 System with Dichotomous Excitation

We now consider the standard Duffing–Holmes with parameter $\epsilon = 0.1$ (Eq. 5.5.1), excited by coin-toss square-wave dichotomous noise $G(t)$

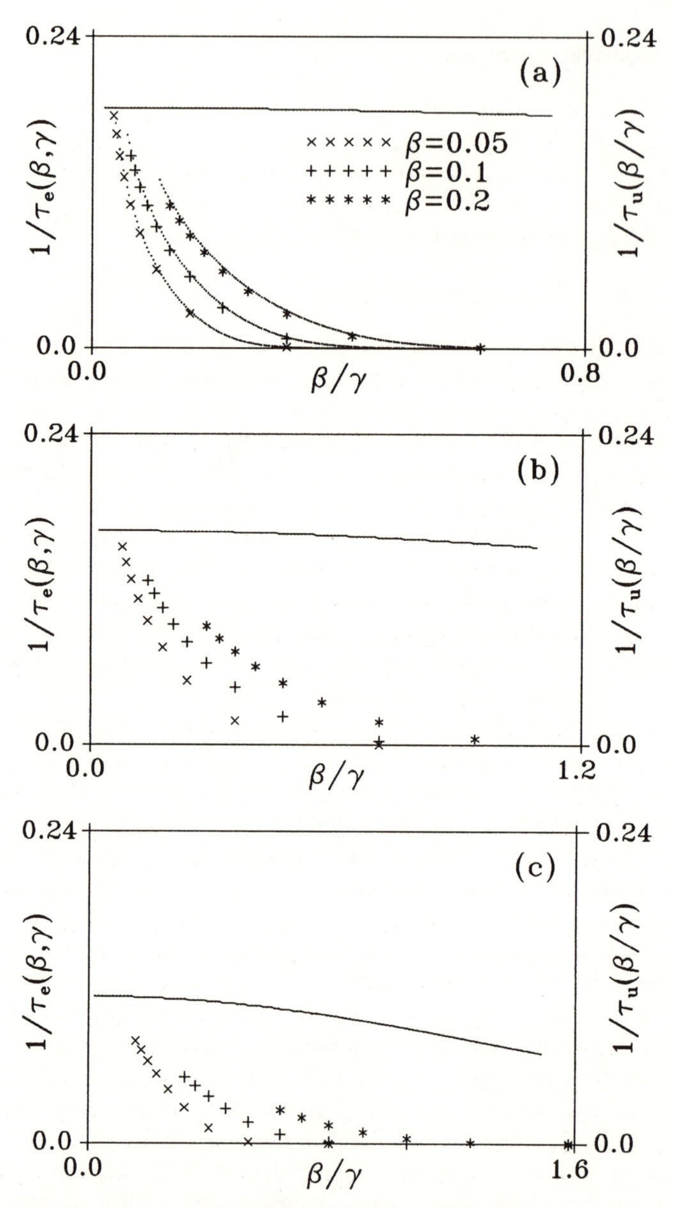

Figure 5.10. Mean zero upcrossing rate $1/\tau_u$ of Melnikov process (solid lines, Eq. 5.5.2) and mean escape rate $1/\tau_\epsilon$ for Duffing–Holmes oscillator ($\epsilon = 1$). (a) White noise excitation (dashed lines, Eq. 5.5.7) and colored noise excitation $c = 0.1$); (b) $c = 0.50$; (c) $c = 2.5$. Note that while $1/\tau_\epsilon$ depends on two parameters (the viscous damping coefficient β and the noise intensity γ), $1/\tau_u$ depends on one parameter, the ratio β/γ.

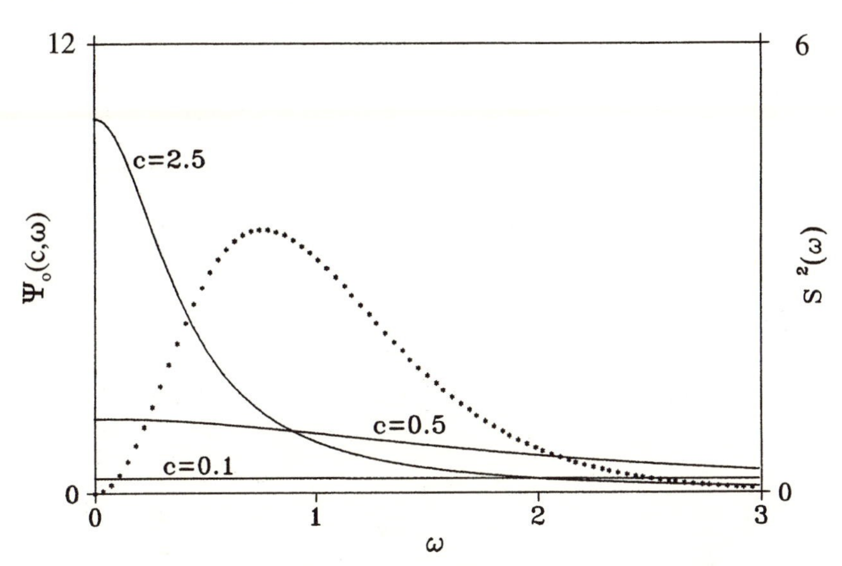

Figure 5.11. One-sided spectral densities of process $G(t)$ with unit variance for three values of c (solid lines), and square of Melnikov scale factor of standard Duffing oscillator for $\gamma = 1$ (dotted line). Note that the proportion of the total power residing at effective frequencies (i.e., at frequencies for which the Melnikov scale factor has relatively large ordinates) is larger for $c = 0.5$ than for $c = 0.1$.

(Eq. 4.2.16). Mean upcrossing rates $1/\tau_u$ for the Melnikov process and mean escape rates $1/\tau_\epsilon$ for that system, obtained by numerical simulation, are shown in Fig. 5.12. Zero upcrossings of the Melnikov process occur if

$$F(t) > 0.9428\beta/\gamma = m$$

(see Eq. 5.4.5), that is, the mean time τ_u between consecutive zero upcrossings for the Melnikov process is equal to the mean time between consecutive upcrossings of the level m for the process $F(t)$.

The rate $1/\tau_u$ is represented by solid lines. As was the case for Gaussian noise excitation, the results show that, for fixed ratio β/γ, the upper bound τ_u^{-1} for the mean escape rate τ_ϵ^{-1} is closer to τ_ϵ^{-1} as the perturbation increases. For example, assume $t_1 = 0.35$, $\beta/\gamma = 0.58$. Figure 5.12 shows that for $\epsilon\beta = 0.05$, $\tau_\epsilon^{-1} \approx 10^{-6}$, while for $\epsilon\beta = 0.25$, $\tau_\epsilon^{-1} \approx 10^{-2}$. For both cases $\tau_u^{-1} \approx 10^{-1}$. For $t_1 = 1.0$, $\epsilon = 0.1$, $\beta = 1.5$, $\beta/\gamma = 0.53$, it is estimated from Fig. 5.8b that $\tau_\epsilon^{-1} \approx 0.03$. This estimate is reasonably consistent with the result of Fig. 5.12c, since the averaging time in Fig. 5.8b is relatively short and, therefore, corresponding estimates of τ_ϵ^{-1} have significant variability.

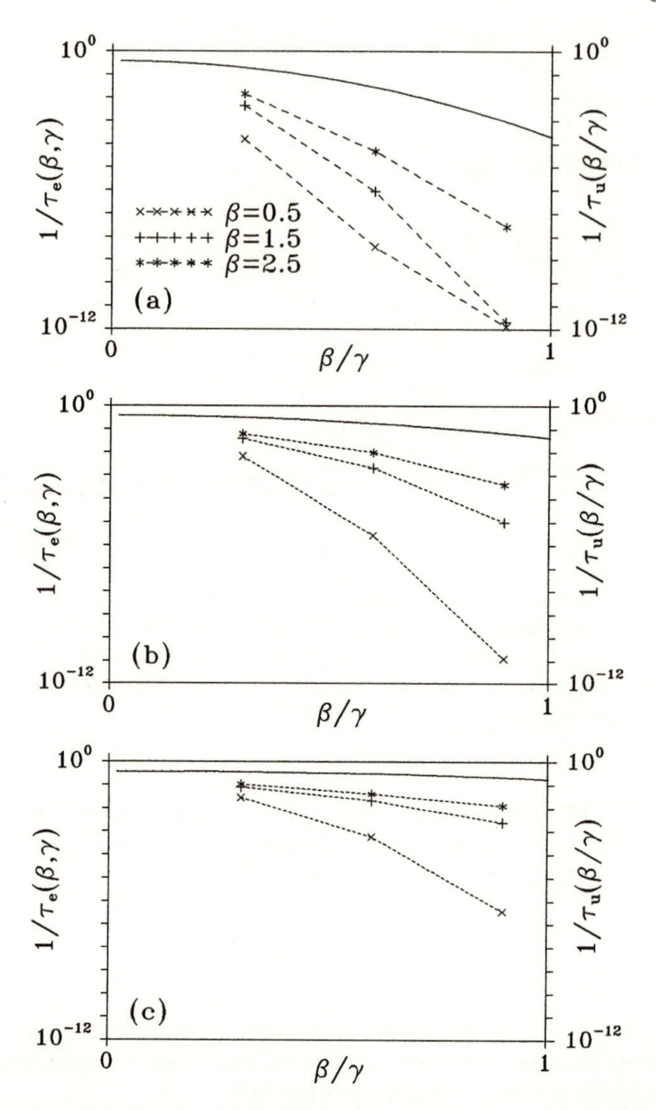

Figure 5.12. Mean zero upcrossing rate $1/\tau_u$ of Melnikov process and mean escape rate $1/\tau_\epsilon$ for Duffing–Holmes oscillator ($\epsilon = 0.1$) excited by dichotomous noise with (a) $t_1 = 0.01$; (b) $t_1 = 0.35$; (c) $t_1 = 1.00$.

5.5.3.3 Melnikov-Based Upper Bound for Probability that Escapes Can Occur

As a numerical example for the upper bound 5.5.5 consider the rf Josephson junction (Eqs. 2.5.20). We assume white noise excitation, $\epsilon = 1$ (Eq. 5.5.1), and parameters $a = 1$, $b_0 = 0$, $\gamma \cong b_1 = 0.025$, $\beta = 0.1$. For these parameters we obtain $\tau_u = 7.66 \times 10^9$ (Eq. 5.5.2) and $\tau_\epsilon = 1.9 \times 10^{2780}$ (Eq. 5.5.7). In this example the discrepancy between τ_u and τ_ϵ is extremely large. However, the lower bound τ_u is itself relatively large, and could therefore be useful if knowledge that $\tau_\epsilon > 10^9$ is of interest for the application at hand. Similarly, consider an interval $T = 10^5$, say. By Eq. 5.5.5, the upper bound to the probability that escapes can occur during that interval is $p_{M,T} = 1 - \exp(-10^{-4}/7.66) \approx 10^{-5}$. The estimated probability that escapes will occur during time $T = 10^5$ is much lower (i.e., $p_{\text{esc},T} = 1 - \exp(-10^{-2775}/1.9) \approx 10^{-2775}$). Though in this case $p_{M,T} \gg p_{\text{esc},T}$, the upper bound estimate $p_{M,T}$ may be sufficiently small to be useful for some applications.

5.6 EFFECTIVE MELNIKOV FREQUENCIES AND MEAN ESCAPE TIME

In this section we discuss the basis for the use of the Melnikov approach to assess the effect of the spectral density of the excitation on the mean escape time τ_ϵ. We illustrate this effect with two examples.

Recall that the spectral density of the Melnikov process is

$$\Psi_M(\omega) = |\boldsymbol{\alpha}(\omega)|^2 \Psi_0(\omega) \tag{5.6.1}$$

(Eq. 5.2.6), where the variance of the excitation $G(t)$ with one-sided spectral density $\Psi_0(\omega)$ is unity, the Melnikov transfer function $\boldsymbol{\alpha}(\omega)$ is defined by Eqs. 4.3.5a, and $|\boldsymbol{\alpha}(\omega)|$ is the Melnikov scale factor. It is convenient to approximate the process $G(t)$ by the process

$$G_N(t) = \sum_{i=1}^{N} a_i \cos(\omega_i t + \phi_{0i}) \tag{5.6.2}$$

(Eq. 4.2.5), where $a_i = [2\Psi_0(\omega_i)\Delta\omega/(2\pi)]^{1/2}$, ϕ_{0i} are uniformly distributed over the interval $[0, 2\pi]$, $\omega_i = i\Delta\omega$, $\Delta\omega = \omega_{\text{cut}}/N$, and ω_{cut} is the frequency beyond which $\Psi_0(\omega)$ vanishes or becomes negligible (the cutoff frequency).

The Melnikov process can be written in the form

$$M(t) = \sum_{i=1}^{N} a_i |\boldsymbol{\alpha}(\omega_i)| \cos(\omega_i t + \phi_{0i} - \Psi_i(\omega_i)) - k \tag{5.6.3}$$

(Eq. 2.5.23). The necessary condition for the occurrence of transitions is that the Melnikov process have simple zeros, that is,

$$\sum_{i=1}^{N} a_i |\boldsymbol{\alpha}(\omega_i)| - k > 0. \tag{5.6.4}$$

Since the arguments of the harmonic functions in Eq. 5.6.3 are incommensurate, Eq. 5.6.4 is approximate; however, the error in (5.6.4) is negligible from the point of view of practical applications. The larger the left-hand side of Eq. 5.6.4, the larger is the strength of the chaotic transport and therefore the larger is the chaotic transition rate.

Equation 5.6.4 can be written

$$\sum_{i=1}^{N}[2\Psi_0(\omega_i)\Delta\omega/(2\pi)]^{1/2}|\boldsymbol{\alpha}(\omega_i)| - k > 0. \qquad (5.6.5)$$

Chaotic transport increases as the sum

$$I = \sum_{i=1}^{N}[\Psi_0(\omega_i)]^{1/2}|\boldsymbol{\alpha}(\omega_i)| \qquad (5.6.6)$$

increases. Therefore, the greatest effectiveness in inducing transitions is achieved by excitations whose frequency components are concentrated at and near the frequency where the Melnikov scale factor $|\boldsymbol{\alpha}(\omega_i)|$ attains its maximum. Comparisons between the effectiveness of excitations with different spectra can be made by inspection: excitations with the same variance but different spectral densities are more or less effective according as their respective frequency content is closer to or farther away from the frequency of the Melnikov scale factor's peak. To the extent that they help or do not help promote transitions, excitation frequencies are referred to as *effective* and *ineffective*, respectively.

Example 5.6.1 *Colored excitation noise spectrum and transitions in the standard Duffing–Holmes oscillator*

The results of Fig. 5.10 were obtained for the standard Duffing–Holmes oscillator, for which the Melnikov scale factor is $|\boldsymbol{\alpha}(\omega)| = S(\omega) = \gamma\sqrt{2}\pi\omega\text{sech}(\pi\omega/2^{1/2})$ and $a = b = 1$ (Eq. 2.5.17). For $\gamma = 1$ (see Remarks at the end of Section 2.5.3 and following Eq. 5.2.6) $S(\omega)$ is shown in Fig. 2.6. We consider the case $\beta = 0.2$, $\beta/\gamma = 0.4$. For $c = 0.5$ (Fig. 5.10b) $\tau_\epsilon \approx 0.06$, while for $c = 0.1$ (Fig. 5.10a) $\tau_\epsilon^{-1} \approx 0.015$. In light of Eq. 5.6.7 this dependence of τ_ϵ on c is explained as follows. The proportion of the total spectral power of the excitation that resides at high frequencies, where the ordinates of the Melnikov scale factor are small, is larger for $c = 0.1$ than for $c = 0.5$ (see Fig. 5.11). Therefore more of the excitation power is ineffective, and the escape rate is smaller for $c = 0.1$ than for $c = 0.5$. The Melnikov scale factor is seen to provide a useful means for assessing qualitatively the relative effect of the spectral density of the excitation upon the rate of escape.

Example 5.6.2 *Effective and ineffective excitations for a quasiperiodically excited system*

We now consider a Duffing–Holmes oscillator with $a = b = 1/2$ (Eq. 2.5.14), $\epsilon\beta = 0.045$, and excitation consisting of a sum of two harmonics. The first harmonic is $\epsilon\gamma G(t) = \lambda \sin(\Omega t)$, $\lambda = 0.114558$, and $\Omega = 0.89$. The second harmonic is $\Delta\lambda \sin(\omega t)$. For $\Delta\lambda = 0$ Monte Carlo simulations yielded a mean escape time $\tau_\epsilon \approx 60$.

Figure 5.13a shows the dependence of the escape rate τ_ϵ on the frequency ω for three values of the amplitude $\Delta\lambda$. Figure 5.13b shows the dependence on frequency ω of the Melnikov scale factor $S(\omega)$ corresponding to $\gamma = 1$.

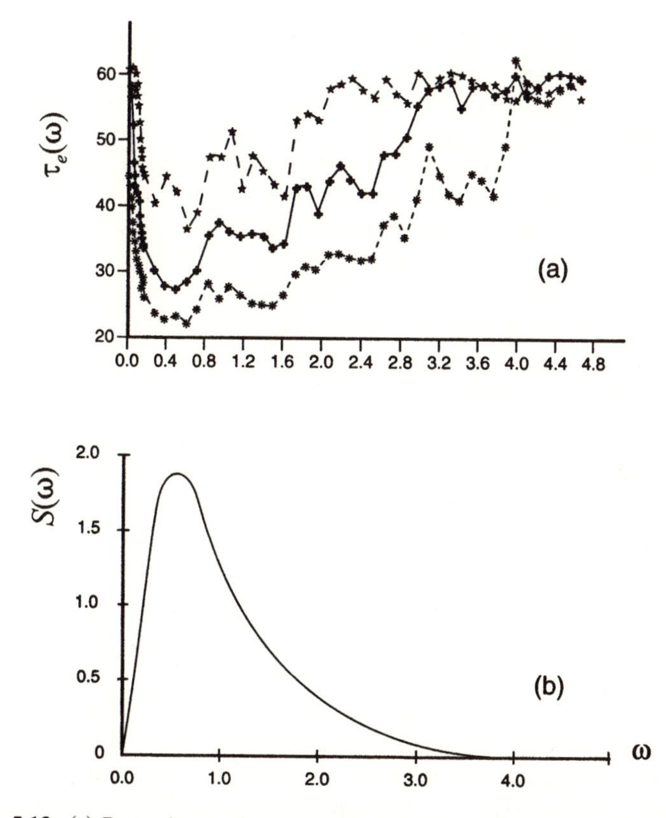

Figure 5.13. (a) Dependence of mean escape time τ_ϵ on frequency ω for three different amplitude $\Delta\lambda$ of the added excitation $\Delta\lambda \sin(\omega t)$. Upper curve: $\Delta\lambda = 0.0005$; middle curve: $\Delta\lambda = 0.004$; lower curve: $\Delta\lambda = 0.032$. (b) Melnikov scale factor (Eq. 2.5.17, $a = b = 1/2$) for $\gamma = 1$. The added excitation is most effective in reducing the mean escape time for frequencies ω close to the frequency of the Melnikov scale factor's peak (after Franaszek and Simiu, 1995).

Figure 5.13 shows that the added excitations $\Delta\lambda \sin(\omega t)$ are most effective in changing τ_ϵ when their frequencies are equal or close to the frequency ω_{pk} of the Melnikov scale factor's peak. Note that the larger $\Delta\lambda$, the larger is the chaotic transport and therefore the shorter is τ_ϵ. It will be seen in Chapters 7, 8, and 12 that the Melnikov-based distinction between effective and ineffective frequencies is useful for designing open-loop control systems that efficiently reduce stochastically induced escape rates, for understanding the effect of the excitation's spectrum on the stochastic resonance phenomenon, and for modeling auditory nerve fiber response.

5.7 SLOWLY VARYING PLANAR SYSTEMS

It was shown in Section 2.7 that, for planar systems to which the Melnikov approach applies, one can define an inner core bounded by a pseudoseparatrix and a region exterior to that core. In general this is not possible for slowly varying planar systems or higher-order systems. Also, for slowly varying planar and higher-order systems, escapes from preferred regions of phase space may occur not only through the mechanism of chaotic transport, but also through other mechanisms to which Melnikov theory is not relevant. The stochastic Melnikov method is applicable only for regions of parameter space for which chaotic transport is the only escape mechanism or, in an approximate manner, for systems whose escape rates associated with alternative mechanisms are significantly lower than those associated with chaotic transport. For an example see Whalen (1996).

5.8 SPECTRUM OF A STOCHASTICALLY FORCED OSCILLATOR: COMPARISON BETWEEN FOKKER-PLANCK AND MELNIKOV-BASED APPROACHES

In this section we consider a chaotic dynamical system for which expressions for the mean time of travel between successive extrema can be found for the case of white noise excitation by using the Fokker-Planck equation, and for the cases of white or colored Gaussian noise by using the Melnikov method. We show that, for white noise, the Fokker-Planck and the Melnikov approach yield expressions of exactly the same form. We also note that the Melnikov approach is much simpler, and that unlike the Fokker-Planck approach it also readily yields expressions for the case of colored noise excitation.

The equations of motion are

$$\dot{x}_1 = x_2, \tag{5.8.1a}$$

$$\dot{x}_2 = x_1 - x_1^3 + \epsilon[\gamma G(t) - (\beta - \alpha x_1^2)x_2] \tag{5.8.1b}$$

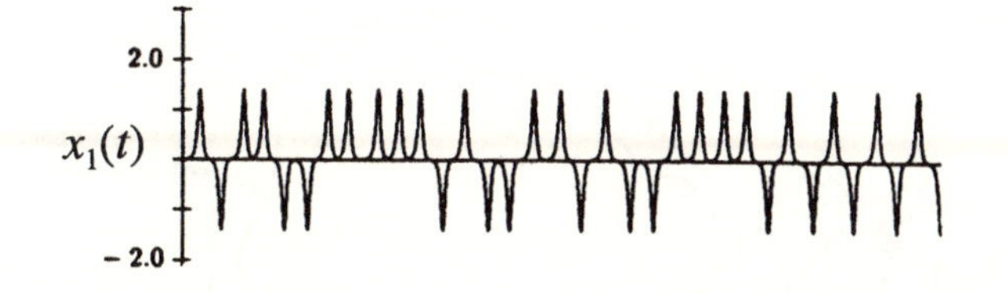

Figure 5.14. Time history of Eq. 5.8.2 (after Brunsden et al. 1989).

($\alpha, \beta > 0, 0 < \epsilon \ll 1$). The system 5.8.1 has a damping term $(\beta - \alpha x_1^2)x_2$ that causes the homoclinic orbit of the unperturbed system to be attracting. This system was studied by Brundsen, Cortell, and Holmes (1989), who adduced theoretical arguments and numerical evidence supporting the assumption that chaotic solutions of Eqs. 5.8.1 may be represented in the form

$$x_1(t) = \sum_{-\infty}^{\infty}(-1)^{a_j} x_{h1}(t - T_j), \tag{5.8.2}$$

where $\{a_j\}$ is a bi-infinite sequence of independent random variables with probabilities $P\{a_j = 0\} = P\{a_j = 1\} = \frac{1}{2}$, $x_{h1}(t)$ is the position coordinate of the unperturbed system's homoclinic orbit (Eq. 2.2.1), and T_j is a random variable marking the time of the jth maximum of $|x_1(t)|$ (Fig. 5.14). The power spectrum of $x_1(t)$ is

$$\Psi_x(f) = (2\pi^2/T)\text{sech}^2(\pi^2 f), \tag{5.8.3}$$

where $T = \langle T_{j+1} - T_j \rangle$ is the mean time between successive maxima of $|x(t)|$.

The mean time T is estimated as follows. We have

$$T = \tau_{\text{out}} + \tau_{\text{in}} \tag{5.8.4}$$

where τ_{out} and τ_{in} are the mean travel times outside and inside U_δ, a square neighborhood of side $2\delta(\epsilon \ll \delta \ll 1)$ of the hyperbolic fixed point (0, 0) (Fig. 5.15). The mean travel time τ_{out} near the attracting homoclinic orbit outside U_δ is, to first order, a constant. Within U_δ the behavior of the solutions is dominated by the solution of the linearized system

$$\dot{u} = \lambda_u u, \qquad \dot{v} = -\lambda_s v, \tag{5.8.5a,b}$$

where the coordinates are parallel to the system's eigenvectors, and the eigenvalues are obtained from the variational equation corresponding to Eqs. 5.8.1.

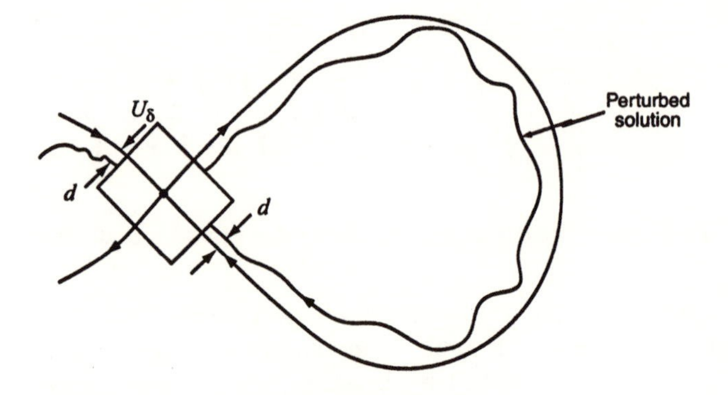

Figure 5.15. Notations.

Integration of Eq. 5.8.5a from $u = d$ to $u = \delta$ yields

$$\tau_{\text{in}} = \lambda_u^{-1} \ln(\delta/d). \tag{5.8.6}$$

The distance d is expressible in terms of the Melnikov distance, and can therefore be written as $d = cd_{\text{mean}}$, where c is a factor of order 1,

$$d_{\text{mean}} \propto \epsilon \times \text{Mean}[|M(t_0)|], \tag{5.8.7}$$

and $M(t_0)$ is the Melnikov function (Brundsen et al. 1989). Therefore

$$\tau_{\text{ln}} = \lambda_u^{-1} \ln\{\delta/[c\epsilon \times \text{Mean}(|M(t_0)|)]\}. \tag{5.8.8}$$

From Eqs. 5.8.4 and 5.8.8,

$$T \approx \tau_{\text{out}} + \lambda_u^{-1} \ln[\delta/(cd_{\text{mean}})] \tag{5.8.9a}$$
$$= K - \lambda_u^{-1} \ln[(\epsilon \times \text{Mean}(|M(t_0)|)]. \tag{5.8.9b}$$

Equations 5.8.9 show that the mean time T depends upon the excitation $G(t)$ through the Melnikov function.

We now specialize Eq. 5.8.9b for the cases of excitation by a harmonic function, colored Gaussian noise, and white Gaussian noise.

For excitation by a *harmonic function*, the Melnikov function can be shown to be

$$M(t_0) = 4\beta/3 - 16\alpha/15 + \gamma\pi\omega\sqrt{2}\,\text{sech}(\pi\omega/2)\sin\omega t_0. \tag{5.8.10}$$

In the particular case $\alpha = 4\beta/5$, Eqs. 5.8.9b and 5.8.10 yield

$$T = K_c - \lambda_u^{-1} \ln[\epsilon\gamma\omega\,\text{sech}(\pi\omega/2)]. \tag{5.8.11}$$

where K_c is a constant.

For excitation by a *colored stationary Gaussian process*, a similar approach and the use of approximations of the type 4.2.1 yield

$$T = K - \lambda_u^{-1} \ln(\epsilon\sigma_M), \tag{5.8.12}$$

where σ_M is the standard deviation of the Melnikov process (Simiu and Frey, 1993b).

In the *white noise* limit (Eqs. 5.2.6, 5.2.7, 2.5.10a, and Eq. 2.5.17 in which $\gamma = 1$)

$$\sigma_M^2 = \gamma^2 \int_0^\infty S^2(\omega)\Psi_0(\omega)d\omega = \gamma^2 \int_0^\infty S^2(\omega)d\omega = 4\pi\gamma^2/3. \tag{5.8.13}$$

Equations 5.8.12 and 5.8.13 yield

$$T = K_1 - \lambda_u^{-1} \ln(\epsilon\gamma) \tag{5.8.14}$$

(Simiu and Frey, 1993b).

For excitation by white noise the following alternative approach was developed by Stone and Holmes (1990). The time τ_{in} is obtained by solving the system written formally as

$$dx = -\lambda_s x dt + \epsilon\gamma dW_x, \tag{5.8.15a}$$

$$dy = \lambda_u y dt + \epsilon\gamma dW_y, \tag{5.8.15b}$$

where dW_x, dW_y are zero-mean, independent Wiener processes[3] and $\epsilon\gamma$ is the noise intensity. The fact that white noise is unbounded would preclude its use for a system with perturbed stable and unstable manifolds, but physical noise processes are bounded and the use of equations whose formal counterparts are Eqs. 5.8.15 is therefore acceptable. Also underlying Eqs. 5.8.15 is the assumption that, when nonlinear terms are ignored, the stable and unstable subsystems evolve independently. By using the Fokker-Planck equation (Stone and Holmes, 1990) for $\alpha = 4\beta/5$ an elaborate procedure not reproduced here yields the result

$$T = K_s - \lambda_u^{-1} \ln(\epsilon\gamma), \tag{5.8.16}$$

[3] A stochastic process $W_0(t)$, $t > 0$, is a Wiener process if $W_0(0) = 0$ and, for each $0 \leq t' < t$, it has independent increments $W_0(t) - W_0(t')$ whose probability density is Gaussian with zero mean and variance $t - t'$.

that is, an expression of exactly the same form as the expression obtained by making use of the expression for the Melnikov process (Eq. 5.8.14). As can be checked by following the details of that procedure, an advantage of the Melnikov-based approach is that it is much simpler and more transparent. A second advantage is that it readily yields the requisite solution not only for white noise excitation but for colored noise as well.

PART 2
Applications

Chapter Six

Vessel Capsizing

This chapter describes an application of the stochastic Melnikov method in naval architecture, based on a model developed by Falzarano et al. (1992) and results obtained by Hsieh, Troesch, and Shaw (1994). In this model the vessel is safe as long as the angle of roll (rotation about the longitudinal axis of the vessel) due to forces induced by random waves does not exceed a critical value.

Section 6.1 describes the model for the vessel dynamics. Section 6.2 presents a numerical example for a specific type of vessel.

6.1 MODEL FOR VESSEL ROLL DYNAMICS IN RANDOM SEAS

The equation of rolling motion of the vessel represents the balance of

- inertia terms, due to (1) the rotational acceleration of the vessel, and (2) the fluid accelerations induced by the vessel's rolling motion (the latter term is empirical)
- empirical damping terms dependent upon the time derivative of the roll angle and representing the energy dissipation in the fluid due to the fluid-vessel interaction
- an empirical restoring term dependent upon the roll angle
- a random forcing term representing the action of random waves[1] on the vessel's sides

Denoting the roll angle by ϕ (Fig. 6.1), the equation of roll motion can be written as

$$(I_{44} + A_{44}(\Omega_0))\phi'' + B_{44}(\Omega_0)\phi' + B_{44q}(\Omega_0)\phi'|\phi'| + \Delta GZ(\phi) = F_{\text{sea}}(\tau) \tag{6.1.1}$$

where I_{44} is the vessel's moment of inertia about the rolling axis, $A_{44}(\Omega_0)$ is the roll hydrodynamic added mass, B_{44} and B_{44q} are the linear and the

[1] Random waves are referred to as seas, while regular waves, which usually persist beyond and after the meteorological event causing them, are referred to as swell.

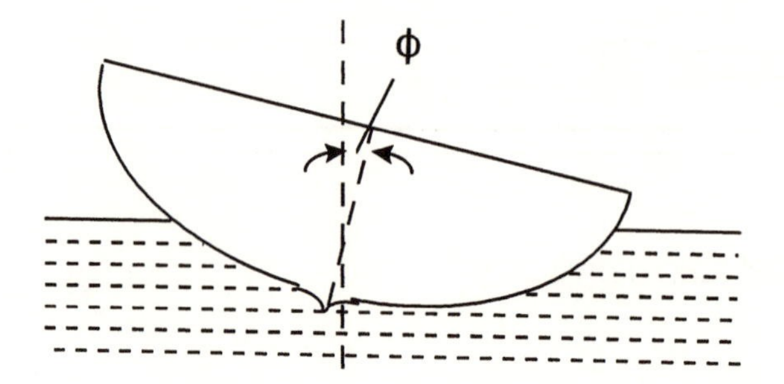

Figure 6.1. A vessel's angle of roll ϕ.

quadratic viscous damping coefficient, respectively, Δ is the vessel displacement[2], $GZ(\phi)$ is an empirical nonlinear rolling restoring moment that depends upon the roll angle and is referred to as the righting curve, $F_{\text{sea}}(\tau)$ is the wave-induced excitation, and a prime denotes differentiation with respect to the time τ. The added mass and damping coefficients are evaluated at a representative frequency Ω_0.

The righting curve is modeled as

$$GZ(\phi) \approx C_1\phi - C_3\phi^3, \tag{6.1.2}$$

where the constants C_1 and C_3 are chosen to match the frequency of small roll oscillations and so that $GZ(\phi^*) = 0$, and ϕ^* is the angle of vanishing stability. In nondimensional form Eq. 6.1.1 is

$$\dot{x}_1 = x_2, \tag{6.1.3a}$$

$$\dot{x}_2 = x_1 + ax_1^3 + \epsilon[-\delta_1 x_2 - \delta_2 x_2|x_2| + f(t)], \tag{6.1.3b}$$

where $x = \phi$, $t = \omega_n\tau$, $\omega_n = \{\Delta C_1/[I_{44} + A_{44}(\Omega_0)]\}^{1/2}$ is the natural frequency of the vessel, $\epsilon\delta_1 = B_{44}(\Omega_0)\omega_n/(\Delta C_1)$, $\epsilon\delta_2 = B_{44q}(\Omega_0)/[I_{44} + A_{44}(\Omega_0)]$, $a = C_3/C_1$, $\epsilon f(t) = F_{\text{sea}}/(\Delta C_1)$, and the dot denotes differentiation with respect to t. For the system 6.1.3 the nondimensional frequency corresponding to a dimensional frequency Ω is $\omega = \Omega/\omega_n$.

The potential associated with the righting curve is shown in Fig. 6.2a. It is M shaped, unlike the potential of the Duffing–Holmes oscillator, which is W shaped (see Fig. 2.1a). The unperturbed system ($\epsilon = 0$) describes undamped rolling in the absence of waves or other forcing. Its phase plane

[2] The displacement is the weight of the water displaced by the vessel.

diagram has a center at the origin, and two saddle points at $(x_1, x_2) = (\pm 1/a^{1/2}, 0)$ connected by the symmetric pair of heteroclinic orbits

$$(x_{h1}(t), x_{h2}(t)) = \{\pm(1/a^{1/2})\tanh(t/\sqrt{2}), \pm[1/(2a)^{1/2}]\text{sech}^2(t/\sqrt{2})\} \qquad (6.1.4)$$

(Fig. 6.2b). The safe region of the phase plane is the region enclosed by the heteroclinic orbits.

The mean of the Melnikov process is obtained from Eqs. 5.2.5 and 2.5.3. (Note that in Eq. 2.5.3 $q_1 = 0$ and $f_1(\mathbf{x}_h) = x_{h2}$; see Eqs. 2.3.1, 2.5.1, and 6.1.3.) The result obtained is

$$E[M(t_0)] = -k = [2\sqrt{2}\delta_1/(3a) + 8\delta_2/(15a^{3/2})]. \qquad (6.1.5)$$

The Melnikov scale factor is obtained by using Eqs. 2.5.10a, 2.5.9, and 2.5.5d in which $\gamma_2 = 1$ (see remark following Eq. 5.2.6). The result is

$$|\alpha(\omega)| = (2/a)^{1/2}[(\pi\omega)/\sinh(\omega\pi/\sqrt{2})]. \qquad (6.1.6)$$

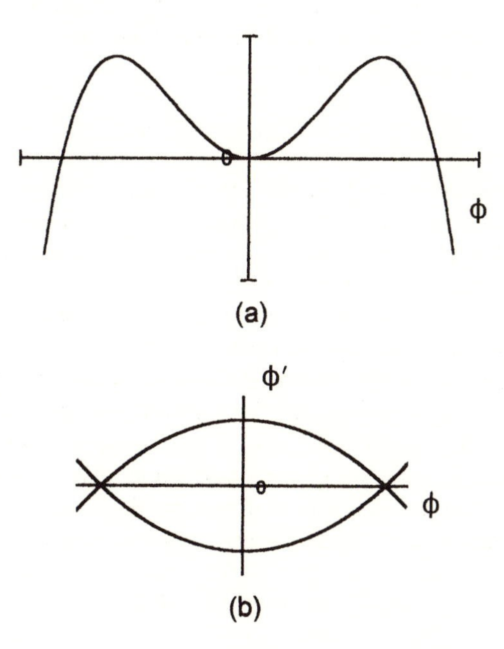

Figure 6.2. Shape of (a) potential well and (b) heteroclinic orbits, for model of rolling vessel.

The spectral density of the Melnikov process is

$$\Psi_M(\omega) = |\alpha(\omega)|^2 \Psi_f(\omega) \qquad (6.1.7)$$

(Eq. 5.2.6), where $\Psi_f(\omega)$ is the spectral density of the wave forces.

6.2 NUMERICAL EXAMPLE

The vessel *Patti-B* has the following characteristics:

> $I_{44} + A_{44}(\Omega_0) = 1.468 \times 10^6$ kg m^2; $B_{44}(\Omega_0) = 3.206 \times 10^3$ kg m^2 s$^-1$;
> $B_{44q}(\Omega_0) = 9.882 \times 10^4$ kg m^2; $\Delta = 2.366 \times 10^6$ N; $C_1 = 0.2138$ m;
> $C_3 = 0.6713$ m

(National Safety Transportation Board, 1979), and a metacentric height of 0.214 m with a corresponding angle of vanishing stability $\phi^* = \pm 32.3°$.

The one-sided spectral density of the incident wave was modeled by the International Ship Structures Congress two-parameter formula

$$\Psi_w(\omega, H_s) = 0.11 H_s^2 [\Omega_c^4/(\omega\omega_n)^5] \exp\{-0.44[\Omega_c/(\omega\omega_n)]^4\}, \qquad (6.2.1)$$

where H_s and Ω_c are the significant wave height[3] and the characteristic wave frequency, respectively. The spectral density of the wave forces is assumed to be given by a simplified, linear model:

$$\Psi_f(\omega) = |F_{\text{roll}}(\omega)|^2 \Psi_w(\omega, H_s). \qquad (6.2.2)$$

The transfer function $F_{\text{roll}}(\omega)$ used in this example was based upon pressures due to waves with small slopes acting on a fixed vessel in the upright position, and is depicted in Fig. 6.3. However, the Melnikov method can be used with more elaborate models of the wave forces.

For $2\pi/\Omega_c = 9$ s, Fig. 6.4 shows the dependence upon significant wave height H_s of the normalized phase space flux $\epsilon\Phi/A_h$, where Φ is the phase space flux factor (Eq. 5.3.3), and A_h is the area enclosed by the heteroclinic orbits. Since $\epsilon\Phi/A_h$ is a measure of the chaotic transport across the pseudoseparatrix of the system 6.1.1, it is also a measure of the probability of escape from the safe area A_h of the phase plane, that is, of the probability of capsizing. Using Eq. 5.3.3 it is shown by Hsieh et al. (1993) that the asymptote to the curve of Fig. 6.4 intersects the H_s axis at

$$H_s^* = -(2\pi)^{1/2} k/(2\sigma_0) \qquad (6.2.3)$$

[3] The significant wave height is defined as the the average height of the highest one-third of the waves.

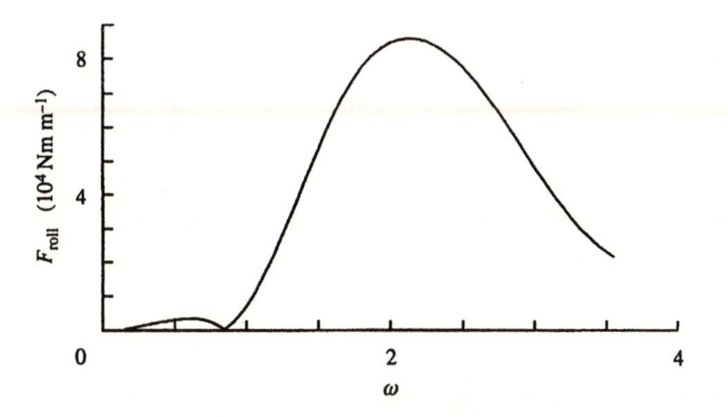

Figure 6.3. Roll moment F_{roll} per unit wave amplitude (after Hsieh, Troesch, and Shaw, 1994).

where σ_0 is the standard deviation of the Melnikov process corresponding to an excitation with unit significant wave height. For $H_s < H_s^*$ the phase space flux is relatively weak; hence, it may be anticipated that the probability of capsizing is relatively small. For this example it was verified by numerical simulations that, for 7.6 m $\leq H_s \leq$ 13.0 m and characteristic wave periods $1\,\text{s} < 2\pi/\Omega_c < 17\,\text{s}$, the probability of capsizing is less than about 0.075 during a time interval of about 35 min, and less than 0.16 during an interval of about 68 min. For this example it is seen that the criterion $H_s < H_s^*$ proposed by Hsieh et al. provides a useful approximate indication of the sea conditions to which correspond low probabilities of capsizing.

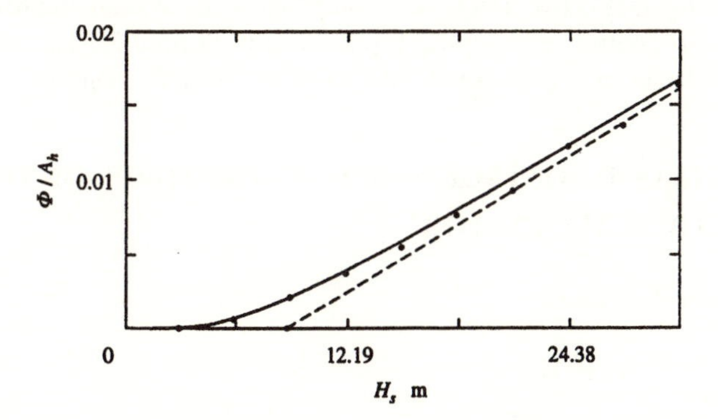

Figure 6.4. Dependence of phase space flux on significant wave height for characteristic wave period $T_c = 9$ s. Solid line, Eq. 5.3.3; dashed line, asymptote; circles, numerical simulation (after Hsieh, Troesch, and Shaw, 1994).

Chapter Seven

Open-Loop Control of Escapes in Stochastically Excited Systems

The performance of certain nonlinear stochastic systems is deemed acceptable if, during a specified time interval, the systems have sufficiently low probabilities of escape from a preferred region of phase space. An example is the motion of a vessel subjected to wave loading (Chapter 6). Given a design sea state, the vessel's motion must have an acceptably small probability of escape from the safe region of phase space. If that probability is too large, its reduction must be achieved by redesigning the system and/or resorting to a suitable control strategy.

In this chapter we describe the basic principle of an open-loop control procedure applicable to systems possessing a Melnikov process. We present two approaches to achieving such a procedure. The approach discussed in Section 7.1 consists of developing a control force which counteracts frequency components of the excitation that are effective in inducing escapes. The selection of those components is determined by the shape of the Melnikov scale factor. The fact that in this approach little energy is wasted for counteracting ineffective frequency components assures an improved efficiency of the control. The approach presented in Section 7.2 uses the phase space flux factor as a measure of the effectiveness of the control.

7.1 OPEN-LOOP CONTROL BASED ON THE SHAPE OF THE MELNIKOV SCALE FACTOR

We consider the system

$$\dot{x}_1 = x_2, \tag{7.1.1a}$$

$$\dot{x}_2 = -V'(x_1) + \epsilon[\gamma G(t) - \beta x_2 - \eta G_c(t)], \tag{7.1.1b}$$

where $\epsilon, \gamma, \beta, \eta > 0$ and ϵ is sufficiently small for Melnikov theory to be valid; $\epsilon\gamma G(t)$ and $-\epsilon\eta G_c(t)$ are the exciting force and the open-loop control force, respectively. We assume $G(t)$ is a stationary, ergodic stochastic process with zero mean, unit variance, and spectral density $\Psi_0(\omega)$.

The mean time τ_u between consecutive zero upcrossings for the system's Melnikov process is a measure of the mean time of escape from a well, τ_ϵ (Section 5.5). The time τ_u—and therefore the time τ_ϵ—due to the exciting force acting alone can be reduced by adding a suitable control force. Assuming a vanishing time lag between the excitation and the control force, a trivial choice of the latter would be $G_c(t) \equiv G(t)$, $\eta < \gamma$. This would decrease the total forcing acting on the system, the mean zero upcrossing rate for the Melnikov process, and the system's mean escape rate. For this trivial choice of the control force, the ratio between the average power of the control force and the average power of the excitation force is $q = \eta^2/\gamma^2$. Our objective is to obtain open-loop control forces that would achieve reductions of the system's mean escape rate comparable to those achieved by the trivial control, but more efficiently, that is, with a decreased ratio q (Simiu and Franaszek, 1997; see also Basios, Bountis, and Nicolis, 1999).

The proposed approach uses the fact that the shape of the modulus of the Melnikov transfer function $|\alpha(\omega)|$ (i.e., the shape of the Melnikov scale factor—see Eq. 2.5.10a) is nonuniform and exponentially decaying. Components of the excitation and of the control force are more effective in affecting escape rates if they correspond to frequency intervals for which $|\alpha(\omega)|$ is relatively large (see Section 5.6). Instead of $G_c(t) \equiv G(t)$, it would be most efficient to apply a control force $G_c(t)$ with as little ineffective frequency content as possible. As indicated earlier, the advantage of this approach over the trivial approach $G_c(t) \equiv G(t)$ is that it can achieve, with less power, a comparable reduction of the system's mean escape rate.

7.1.1 Dynamical System and Excitation Spectra

For specificity we consider the standard Duffing–Holmes oscillator (i.e., the potential in Eq. 7.1.1 is given by Eq. 2.5.14c in which $a = b = 1$) with parameters $\epsilon = 0.1$ and $\beta = 0.45$. The Melnikov scale factor is then given by Eq. 2.5.17 (for this oscillator $|\alpha(\omega)| \equiv S(\omega)$). We examine the case of the excitation $G(t)$ with spectral density

$$\Psi_0(\omega) = \begin{cases} 0.03990\ln(\omega) + 0.12829, & 0.04 \le \omega \le 0.4 \\ 0.05755\ln(\omega) + 0.14493, & 0.4 \le \omega \le 1.2 \\ -0.38301[\ln(\omega)]^2 + 1.06192\ln(\omega) & \\ \quad -0.02941, & 1.2 \le \omega \le 15.4 \end{cases} \tag{7.1.2}$$

(Fig. 7.1). To a first approximation Eq. 7.1.2 is a rescaled spectral representation of low-frequency fluctuations of the horizontal wind speed (van der Hoven, 1957). The square of the Melnikov scale factor, $S^2(\omega)$, and the spectral density of the Melnikov process, $\Psi_0(\omega)S^2(\omega)$ (see Section 5.6) are displayed in Figs. 7.2a and 7.2b, respectively, for $\gamma = 1$. Figure 7.2 shows

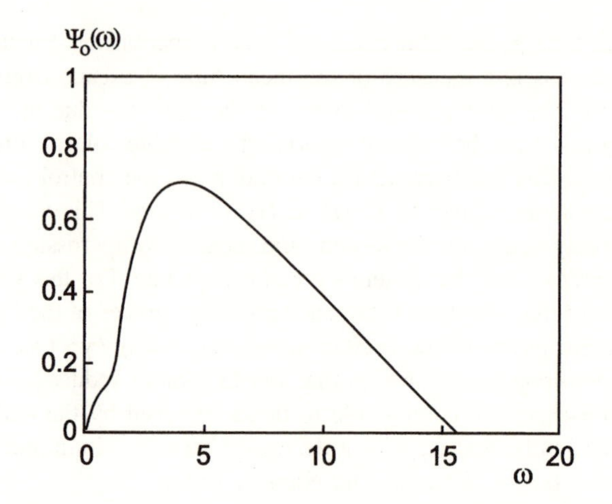

Figure 7.1. Spectral density of stochastic process $G(t)$.

that, owing to the shape of $S^2(\omega)$, only some of the frequency components of the excitation $G(t)$ contribute significantly to the spectral density of the uncontrolled system's Melnikov process. It is only those components that need to be counteracted by a control force. Components with frequencies $\omega > 4$, which are in effect suppressed, and components with frequencies $0 \leq \omega < 0.3$ and $2.5 < \omega \leq 4$, which are very strongly reduced, are ineffective by virtue of the properties of the system and need not be counteracted by the control force.

7.1.2 Control Forces

The control process $G_c(t)$ must be a function of the excitation process $G(t)$. For the trivial control case mentioned earlier this function is simply $G_c(t) \equiv G(t)$. In practice, however, a control force reproducing the excitation process $G(t)$ cannot be achieved instantaneously. The expression of the control process must therefore reflect the existence of a time lag ℓ. In addition, a filter referred to as a Melnikov effectiveness filter is used that causes the spectrum of the control process to have fewer or smaller components with ineffective frequencies than the spectrum of the excitation, since such components are of little or no help in reducing escapes. We also assume the use of an additional filter, referred to as the primary filter, that further modifies the spectrum of the control process. We therefore represent the control process in the form

$$G_c(t) = [\lambda_{\epsilon} * \delta_{t-\ell} * \lambda_p * G](t) \tag{7.1.3}$$

where the asterisk denotes convolution (see Eqs. 5.2.4), $\lambda_p(t)$, $\delta_{t-\ell}(t-\ell)$, and $\lambda_\epsilon(t)$ are the impulse response functions of the primary filter, time lag

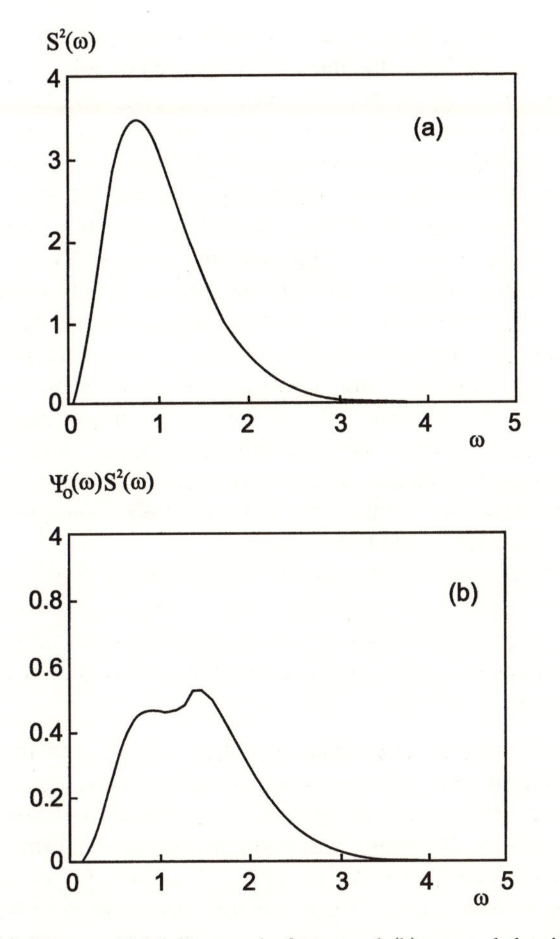

Figure 7.2. (a) Square of Melnikov scale factor and (b) spectral density of uncontrolled system's Melnikov process, for $\gamma = 1$.

filter, and Melnikov effectiveness filter, respectively, and $\delta_{t-\ell}$ is the Dirac delta function. The transfer functions of the primary, time lag, and Melnikov effectiveness filters are $\Lambda_p(\omega)$, $\Lambda_\ell(\omega) = \exp(-j\omega\ell)$, and $\Lambda_\epsilon(\omega)$, respectively. The primary and Melnikov effectiveness filters play complementary roles in achieving a desirable spectral density of the control process.

We consider four types of control, chosen to make it possible to assess the benefits of reducing the ineffective frequency components through the use of the Melnikov effectiveness filter.

(a) Control type a, or trivial control with time lag ℓ, seeks to counteract the excitation by applying a control force proportional and of opposite sign to $\epsilon\gamma G(t)$, that is, $|\Lambda_p(\omega)| \equiv 1$, $\Lambda_\ell(\omega) = \exp(-j\omega\ell)$, and $|\Lambda_\epsilon(\omega)| \equiv 1$. The smaller the lag ℓ, the more effective the control.

(b) Control type b, or control with time lag ℓ and Melnikov effectiveness filter, differs from and is more efficient than control type a insofar as it utilizes the information provided by the Melnikov scale factor $|\alpha(\omega)| \equiv S(\omega)$ as follows. As noted earlier, Figs. 7.1 and 7.2 show that, owing to their suppression by $S^2(\omega)$, spectral components with frequencies $0 \leq \omega < \omega_1$, where $\omega_1 < 0.3$, say, and frequencies $\omega > \omega_2$, where $\omega_2 = 2.5$, say, contribute little to the spectral density of the Melnikov process. In other words the spectral components inside (outside) the interval (ω_1, ω_2) are effective (ineffective). For control type b, just as for control a, $|\Lambda_p(\omega)| \equiv 1$ and $\Lambda_\ell(\omega) = \exp(-j\omega\ell)$. However, $|\Lambda_\epsilon(\omega)| = 1$ only over the frequency interval where the components are effective; over the interval where they are ineffective $|\Lambda_\epsilon(\omega)| = 0$.

(c) Control type c is similar to control type a, except that $|\Lambda_p(\omega)| \neq 1$. In our simulations we use the filter with impulse response function shown in Fig. 7.3, with $A = 0.1$, $B = 2.25$.

(d) Control type d is similar to control type c, except that $|\Lambda_\epsilon(\omega)| = 1$ for the interval where the frequency components are effective; outside that interval $|\Lambda_\epsilon(\omega)| = 0$; that is, control type d has the same relationship with respect to control type c as control type b does with respect to control type a.

In all simulations we assumed a time lag $\ell = 0.1$. The frequencies ω_1 and ω_2 defining the intervals over which inefficient components were filtered out in (b) and (d) were chosen by examining the spectral densities of the Melnikov processes. The choice was consistent with the information inherent in the Melnikov scale factor, that is, $\omega_1 = 0.3$, $\omega_2 = 2.5$.

The strength of the control force was assumed to be $\eta = 0.5\gamma$ for control types b and d. For control types a and c η was chosen so that the

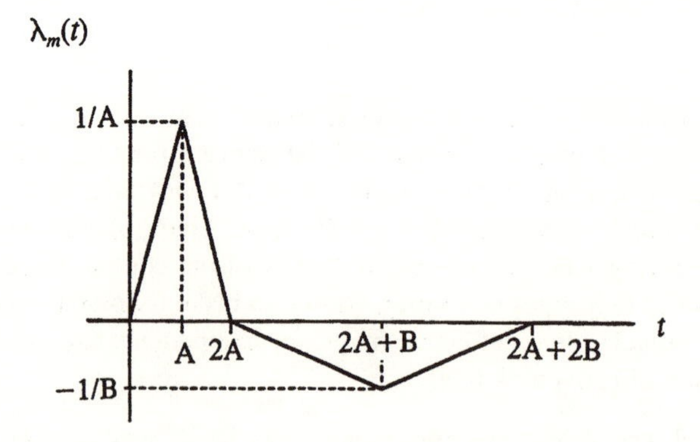

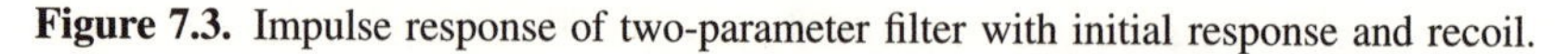

Figure 7.3. Impulse response of two-parameter filter with initial response and recoil.

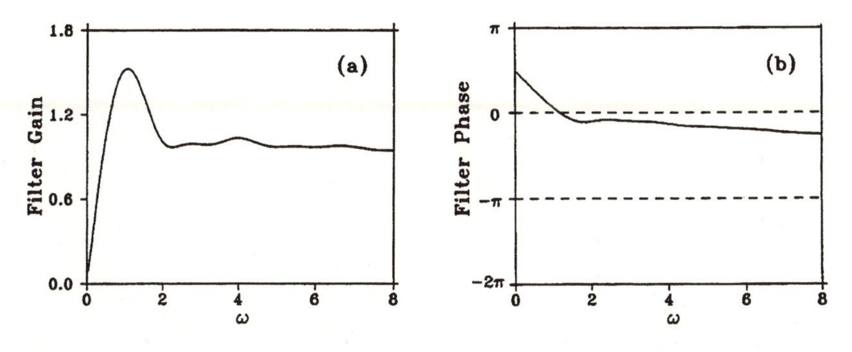

Figure 7.4. (a) Gain and (b) phase angle, for filter of Fig. 7.3 with $A = 0.1$ and $B = 2.25$.

control forces have the same average power (i.e., the same variance) as the control forces for types b and d, respectively. This choice yielded $\eta = 0.167\gamma$ for control type a and $\eta = 0.197\gamma$ for control type c.

The primary filter of Fig. 7.3 has transfer function $\Lambda_p(\omega) = R(\omega) + jI(\omega)$, where

$$R(\omega) = r^2(A\omega/2)\cos(A\omega) - r^2(B\omega/2)\cos(2A + B)\omega, \quad (7.1.4a)$$

$$I(\omega) = -r^2(A\omega/2)\sin(A\omega) + r^2(B\omega/2)\sin(2A + B)\omega, \quad (7.1.4b)$$

and $r(\omega) = \sin(\omega)/\omega$. Equations 7.1.4 were obtained from expressions available, e.g., in Papoulis (1962). We show in Fig. 7.4 the dependence on frequency of the filter gain $[R^2(\omega) + I^2(\omega)]^{1/2}$ and phase $\tan^{-1}[I(\omega)/R(\omega)]$.

7.1.3 Results of Numerical Simulations

Results of numerical simulations are shown in Fig. 7.5. We compare mean escape rates induced by control forces modified to take advantage of the system's Melnikov properties (i.e., forces type b and d) with their unmodified counterparts (forces type a and c, respectively). Recall that the control forces corresponding to curves b and d in Fig. 7.5 have the same average power as those corresponding to curves a and c, respectively, and that they were obtained from the latter by using Melnikov theory to eliminate inefficient components.

For example, Fig. 7.5 shows that, given the external excitation $\epsilon\gamma = 0.15$, the escape rate reduction due to the use of a control force type b is about 20 times stronger than that due to a control force type a having the same average power; while control force type d is about five times more effective than control force type c with the same power. Note that the effectiveness of the control force increases as $\epsilon\gamma$ decreases.

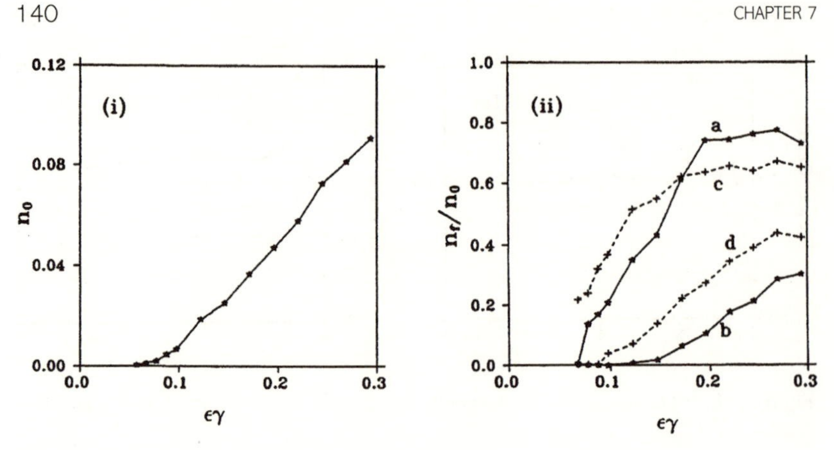

Figure 7.5. (i) Escape rate n_0 for uncontrolled system; (ii) ratio n_f/n_0 between escape rates of system with control types a,b,c,d on the one hand and of uncontrolled system on the other. Note that for excitation $\epsilon\gamma = 0.15$ the escape rate for the Melnikov-based control type b is about 20 times lower than for control type a with same power. This advantage decreases as $\epsilon\gamma$ increases (after Simiu and Franaszek, 1991).

The numerical simulations presented in Fig. 7.5 suggest that controls based on the information contained in the Melnikov transfer function can help to stabilize efficiently a system subjected to random excitation. The degree to which an efficient Melnikov-based open-loop control can be accomplished in practice depends upon the system under consideration (i.e., its Melnikov characteristics), the spectral density of the excitation, the magnitude of the time lag, and the characteristics of the filters used to obtain the control force.

7.2 PHASE SPACE FLUX APPROACH TO CONTROL OF ESCAPES INDUCED BY STOCHASTIC EXCITATION

Having established that the Melnikov approach to open-loop control can be useful, we now consider the use of the phase space flux factor—a functional dependent upon the characteristics of the Melnikov process—as a control objective. A decrease of the flux factor achieved by a control force reflects a reduction of the chaotic transport across the pseudoseparatrix and, therefore, a reduction of the system's mean escape rate (see, for example, Fig. 5.6b).

Our aim is to derive an index reflecting the reduction of the flux factor in terms of the system, forcing, and filter characteristics (Frey and Simiu, 1996). For simplicity we assume $|\Lambda_\epsilon(\omega)| \equiv 1$ and that the primary filter has a transfer function $\Lambda_p(\omega)$ (see Section 7.1.2) capable of enhancing effective components and reducing or suppressing ineffective ones. We consider the system 7.1.1 with standard Duffing potential and control force 7.1.3.

7.2.1 Condition for System Stabilization by Control Force

Recall that the phase space flux factor as defined in Sections 2.7 and 5.3 is proportional to first order to the average area under the curve defining the positive part of the Melnikov process in a phase space slice. We denote by $\gamma M(t)$ and $\eta M_c(t)$ the zero-mean fluctuating parts of the Melnikov process induced by the excitation $\epsilon\gamma G(t)$ and the control $\epsilon\eta G_c(t)$, respectively. The flux factor is

$$\Phi = E[(\gamma M(t) - \eta M_c(t) - k)^+]. \tag{7.2.1}$$

Define the random variable $\sigma Z(t) = \gamma M(t) - \eta M_c(t)$, where $Z(t)$ has unit variance and cumulative distribution function $F_z(z)$. The variance of the process $\sigma Z(t)$ is

$$\sigma^2 = \mathrm{Var}[\gamma M(t) - \eta M_c(t)] = \gamma^2\sigma_M^2 + \eta^2\sigma_{M_c}^2 - 2\gamma\eta\sigma_{MM_c}, \tag{7.2.2}$$

where σ_M^2, $\sigma_{M_c}^2$, and σ_{MM_c} are the variances and the covariance of the processes $M(t)$ and $M_c(t)$. The flux factor can be written as

$$\begin{aligned}\Phi = E[(\sigma Z - k)^+] &= \int_{k/\sigma}^{\infty} (\sigma z - k)\, dF_z(z) = \sigma \int_{k/\sigma}^{\infty} z\, dF_z(z) \\ &- k \int_{k/\sigma}^{\infty} dF_z(z) = \sigma \int_{k/\sigma}^{\infty} [1 - F_z(z)]\, dz,\end{aligned} \tag{7.2.3}$$

the last result of Eq. 7.2.3 is obtained by noting that

$$\int z\, dF_z(z) = -z[1 - F_z(z)] + \int [1 - F_z(z)]\, dz$$

and that for distributions of interest $\lim_{z\to\infty} z[1 - F_z(z)] = 0$. In the particular case of Gaussian excitation and control force, Z is also Gaussian and the flux factor Φ is given by Eq. 5.3.3 in which we substitute σ for σ_M.

From Eq. 7.2.3 it follows that for $\sigma = 0$, $\Phi = 0$. It can be verified by the application of the chain rule to Eq. 7.2.3 that, if the probability density function $f_z(z)$ decreases faster than $1/z^3$ as $z \to \infty$, then $d\Phi/d\sigma = 0$ and $d^2\Phi/d\sigma^2 = 0$, and that, for $\sigma > 0$, $d\Phi/d\sigma > 0$ and $d^2\Phi/d\sigma^2 > 0$. The curve $\Phi(\sigma)$ is therefore similar to the curve of Fig. 5.6a. Since Φ decreases monotonically as σ decreases, Eq. 7.2.2 shows that the flux factor is smaller for the controlled system than for the system with no control force ($\eta = 0$) if and only if $\sigma < \gamma\sigma_M$, or

$$\eta/\gamma < 2\sigma_{MM_c}/\sigma_{M_c}^2. \tag{7.2.4}$$

The optimal ratio η/γ, that is, the ratio that, given γ, $M(t)$, and $M_c(t)$, minimizes σ^2, is obtained by equating to zero the derivative of σ^2 (Eq. 7.2.2) with respect to η. This yields

$$\eta_{\text{opt}}/\gamma = \sigma_{MM_c}/\sigma^2_{M_c}, \tag{7.2.5}$$

and, after substitution of η_{opt} for η in Eq. 7.2.2,

$$\sigma^2_{\text{opt}} = \gamma^2\sigma^2_M\{1 - [\sigma_{MM_c}/(\sigma_M\sigma_{M_c})]^2\} = \gamma^2\sigma^2_M(1 - \rho^2_{MM_c}),$$

where $0 < \rho_{MM_c} = \sigma_{MM_c}/(\sigma_M\sigma_{M_c}) \leq 1$ is the correlation coefficient of the processes M, M_c (see Section 4.1.6). We denote

$$Q = (1 - \rho^2_{MM_c})^{1/2}, \tag{7.2.6}$$

$0 \leq Q < 1$, and refer to it as the flux reduction index. Since, as was shown earlier, the flux factor decreases monotonically as σ decreases, the smaller the index Q, the more effective is the control force.

7.2.2 Effectiveness of Control as a Function of System, Excitation, and Control Filter Characteristics

Expressions for σ^2_M, $\sigma^2_{M_c}$ and σ_{MM_c} as functions of system, excitation, and control filter characteristics can be obtained as follows. We have

$$\sigma^2_M = (1/2\pi)\int_0^\infty |\alpha(\omega)|^2\Psi_0(\omega)\,d\omega \equiv J_0 \tag{7.2.7}$$

where $\Psi_0(\omega)$ is the one-sided spectral density of the process $G(t)$.

The control force is obtained from the excitation force by Eq. 7.1.3. Since we assume $|\Lambda_\epsilon(t)| \equiv 1$, the variance of the process $M_c(t)$ is

$$\sigma^2_{M_c} = (1/2\pi)\int_0^\infty |\Lambda_p(\omega)|^2|\alpha(\omega)|^2\Psi_0(\omega)\,d\omega \equiv J_1. \tag{7.2.8}$$

The covariance of the processes $M(t)$, $M_c(t)$ is

$$\sigma_{MM_c} = (1/2\pi)\int_0^\infty |\alpha(\omega)|^2\Psi_0(\omega)\Lambda_p(\omega)e^{-j\omega\ell}\,d\omega. \tag{7.2.9}$$

For large ω, $|\alpha(\omega)|$ is small and the integrand is negligibly small. We may therefore disregard large values of ω and, if the lag ℓ is small, we may assume $\cos(\omega\ell) \approx 1$, $\sin(\omega\ell) \approx \omega\ell$. We then have

$$\sigma_{MM_c} \approx (1/2\pi)\left[\int_0^\infty |\alpha(\omega)|^2\Psi_0(\omega)R(\omega)\,d\omega + \ell\int_0^\infty |\alpha(\omega)|^2 \times \Psi_0(\omega)I(\omega)\omega\,d\omega\right] \equiv J_2 - J_3\ell. \tag{7.2.10}$$

Therefore, for small lag ℓ, Eq. 7.2.4 becomes

$$J_1 \eta/(2\gamma) + J_3 \ell < J_2. \tag{7.2.11}$$

From Eq. 7.2.5, the optimal control strength is

$$\eta_{\text{opt}} = \gamma(J_2 - J_3 \ell)/J_1. \tag{7.2.12}$$

The flux reduction index is

$$Q = [1 - (J_2 - \ell J_3)^2/(J_0 J_1)]^{1/2}. \tag{7.2.13}$$

Through the integrals J_0, J_1, J_2, and J_3, Q depends upon the Melnikov characteristics of the dynamical system, the spectral density of the excitation process, and the characteristics of the control filter. For the Duffing–Holmes oscillator with parameters $a = b = 1$, spectral density of the excitation $\Psi_0(\omega) = 2\pi/5$ for $0 < \omega < 5$ and $\Psi_0(\omega) = 0$ elsewhere, control filter with the impulse response function of Fig. 7.3, and time lag $\ell = 0$, the dependence of Q on the filter parameters A and B is shown in Fig. 7.6.

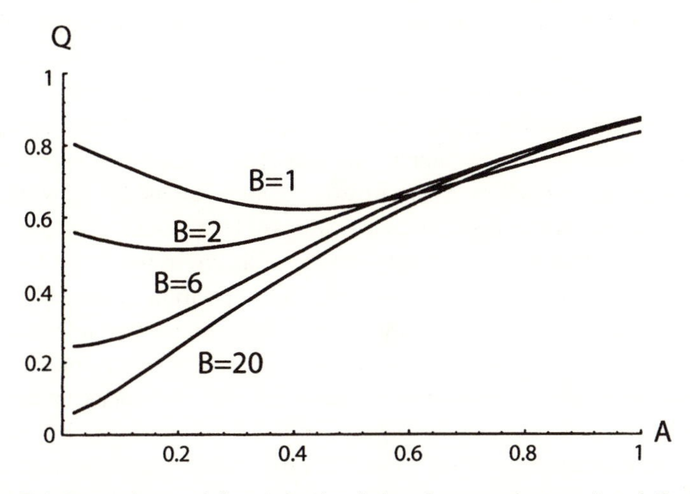

Figure 7.6. Dependence of flux reduction index Q on parameters A and B of filter with impulse response function in Fig. 7.3 (after Frey and Simiu, 1996).

Chapter Eight

Stochastic Resonance

In this chapter we briefly review the stochastic resonance phenomenon, for which we provide an interpretation in chaotic dynamics terms. We use the Melnikov method to (1) assess the role of the noise spectrum in stochastic resonance, and (2) extend the definition of stochastic resonance, that is, show that stochastic resonance can be induced not only by increasing the stochastic excitation of the system but, alternatively, by adding a deterministic excitation. We present results of numerical simulations which show the usefulness of the Melnikov approach to stochastic resonance.

In Section 8.1 we define stochastic resonance and briefly describe its underlying mechanism. In Section 8.2 we describe a typical dynamical system chosen to illustrate our approach, and state the corresponding Melnikov necessary condition for chaos, which we will use as a tool for investigating stochastic resonance. In Section 8.3 we consider the case of a bistable *deterministic* system excited by a sum of two harmonic terms. We show that, for this excitation, chaotic behavior makes it possible to define an output signal-to-noise ratio, that is, the ratio between the ordinates at the signal frequency of the peak output spectrum and the broadband spectrum associated with the chaotic behavior. We show how Melnikov theory can be used to enhance the signal-to-noise ratio in this case. In Section 8.4 we discuss *classical stochastic resonance*, that is, stochastic resonance wherein the enhancement of the signal-to-noise ratio is achieved by adding noise to the system. For classical stochastic resonance we show how Melnikov theory can be used to assess the effect of the shape of the noise spectrum on signal-to-noise enhancement. In Section 8.5 we show that, for a system with signal and noise, the output signal-to-noise ratio can be increased more effectively by adding to the system a harmonic excitation selected in accordance with Melnikov theory, rather than by increasing the noise, as is the case in classical stochastic resonance. Based on this method, we describe in Section 8.6 the principle of a proposed nonlinear transducing device for enhancing signal-to-noise ratio.

8.1 DEFINITION AND UNDERLYING PHYSICAL MECHANISM OF STOCHASTIC RESONANCE. APPLICATION OF THE MELNIKOV APPROACH

For a class of multistable systems excited by a periodic signal and noise, the improvement of the signal-to-noise ratio achieved by the apparently paradoxical means of *increasing* the noise intensity is known as stochastic resonance (or, as it will be referred to here, classical stochastic resonance) (Moss, Pierson, and O'Gorman, 1994).

The essence of the physical mechanism underlying classical stochastic resonance can be described as follows (McNamara and Wiesenfeld, 1989). Consider the motion in a double-well potential of a lightly damped particle subjected to (a) a harmonic excitation (i.e., a signal) with low frequency ω_0, and (b) a stochastic excitation. The harmonic excitation is assumed to have small enough amplitude that, by itself (i.e., in the absence of the stochastic excitation), it is unable to move the particle from one well to another. However, under the combined action of the harmonic excitation *and* the stochastic excitation, escapes do occur, and the spectral density of the output has a broadband portion. We denote the characteristic rate, that is, the mean escape rate from a well under that combined action, by $\alpha = 2\pi n/T$, where n is the total number of exits from a well during a sufficiently long time interval T.

We consider the behavior of the system as we increase the noise while the amplitude and frequency of the harmonic excitation are unchanged. For zero noise excitation, $\alpha = 0$, as noted earlier. For very small noise excitation $\alpha < \omega_0$. As the noise excitation increases, the ordinate of the broadband spectral density at the frequency ω_0, denoted $P_n(\omega_0)$, increases, as does the characteristic rate α. Experimental and numerical studies show that, for $\alpha \approx \omega_0$, there occurs a cooperative effect (i.e., a synchronization-like phenomenon, as it is referred to by Shulgin, Neiman, and Anishchenko, 1995) wherein the signal output spectrum $P_s(\omega_0)$ increases as the noise intensity increases. Remarkably, the increase of $P_s(\omega_0)$ with noise is faster than that of $P_n(\omega_0)$. This results in an enhancement of the signal-to-noise ratio, as illustrated later in this section. The synchronization-like phenomenon plays a key role in the mechanism just described.

Melnikov theory yields qualitative results on the basis of which useful inferences can be made about the behavior of systems exhibiting stochastic resonance. Basically, we use the following facts: (1) for a wide class of systems, deterministic and stochastic excitations play qualitatively equivalent roles in inducing chaotic motions with escapes over a potential barrier (see Fig. 5.1), and (2) in both the deterministic and stochastic case the motions are characterized by broadband spectra associated with chaotic behavior (for the deterministic case see, e.g., the end of Section 3.7). The system's characteristic rate can be increased—and the synchronization-like phenomenon that results in an improved signal-to-noise ratio can be caused—either by increasing the stochastic excitation or by adding a deterministic excitation. Also, since Melnikov theory provides information on excitation frequencies that are effective in increasing a system's characteristic rate, a Melnikov approach makes it possible to assess simply and elegantly the role of the excitation's spectral shape in the enhancement of the signal-to-noise ratio, a problem of interest in classical stochastic resonance (Hänggi et al., 1993).

8.2 DYNAMICAL SYSTEMS AND MELNIKOV NECESSARY CONDITION FOR CHAOS

We consider second-order dynamical systems described by the equation

$$\ddot{x} = -V'(x) - \beta\dot{x} + F(t) \tag{8.2.1}$$

where $V(x)$ is a potential function, $\beta > 0$, and the excitation $F(t)$ is specified below. The unperturbed counterpart of Eq. 8.2.1 is the Hamiltonian system

$$\ddot{x} = -V'(x). \tag{8.2.2}$$

We assume Eq. 8.2.2 has a hyperbolic fixed point connected to itself by a homoclinic orbit or two hyperbolic fixed points connected by heteroclinic orbits. For specificity we consider the standard Duffing–Holmes equation with the double-well potential $V(x)$ given by Eq. 2.5.14c in which $a = b = 1$.

First, suppose that the excitation is harmonic, that is, $F(t) = A_0 \cos(\omega_0 t)$. The necessary condition for chaos is the Melnikov inequality

$$-4\beta/3 + A_0 S(\omega_0) > 0 \tag{8.2.3}$$

where $S(\omega) = \sqrt{2}\pi\omega \operatorname{sech}(\pi\omega/2)$ is the Melnikov scale factor corresponding to a harmonic forcing with amplitude unity (i.e., $\gamma = 1$; see Eq. 2.5.17 and remark following Eq. 2.5.21).

Next, suppose that

$$F(t) = A_0 \sin(\omega_0 t + \phi_0) + A_a \sin(\omega_a t) + \sum_{k=1}^{K} a_k \sin(\omega_k t + \phi_k). \quad (8.2.4)$$

For this case the Melnikov necessary condition for chaos is

$$-4\beta/3 + A_0 S(\omega_0) + A_a S(\omega_a) + \sum_{k=1}^{K} a_k S(\omega_k) > 0, \quad (8.2.5)$$

where $S(\omega)$ has the same expression as for Eq. 8.2.3. (Equation 8.2.5 follows from Eq. 2.5.23a and the fact that, for the Duffing–Holmes oscillator, $C(\omega) \equiv 0$; see Example 2.5.1.)

Finally, suppose that the system excitation is

$$F(t) = A_0 \sin(\omega_0 t + \phi_0) + A_a \sin(\omega_a t) + \gamma G(t), \quad (8.2.6)$$

where $G(t)$ is a Gaussian process with unit variance and one-sided spectral density $\Psi_0(\omega)$. Over any finite time interval, however large, each realization of the process may be approximated as closely as desired by a sum

$$G_N(t) = \sum_{k=1}^{K} b_k \sin(\omega_k t + \phi_k), \quad (8.2.7)$$

so that the Melnikov inequality, that is, the Melnikov necessary condition for chaos, can be written as Eq. 8.2.5, in which $a_k = \gamma b_k$. In Eq. 8.2.7, $b_k = (2\Psi_0(\omega)\Delta\omega/(2\pi))^{1/2}$, ϕ_k are random phase angles with uniform distribution in the interval $[0, 2\pi]$, $\omega_k = k\Delta\omega$, $\Delta\omega = \omega_{\text{cut}}/K$, and ω_{cut} is the frequency beyond which the spectrum of $G(t)$ vanishes or is negligibly small (Eq. 4.2.5).

For the damped, forced system, the strength of the chaotic transport, and therefore the characteristic rate α, increase as the left-hand side of the Melnikov necessary condition for chaos becomes larger. This is true regardless of whether the excitation is deterministic or stochastic. Moreover, as was mentioned in Section 8.1, and again regardless of whether the excitation is deterministic or stochastic, a qualitative feature of any chaotic motion is that its spectral density has a broadband portion with significant energy content at and near the frequency equal to the system's characteristic rate α.

8.3 SIGNAL-TO-NOISE RATIO ENHANCEMENT FOR A BISTABLE DETERMINISTIC SYSTEM

To show that stochastic resonance is a phenomenon of a chaotic dynamical nature we consider a strictly deterministic case where the excitation is a sum of a harmonic signal and an added harmonic, that is, in Eq. 8.2.1 $F(t) = A_0 \sin(\omega_0 t) + A_a \sin(\omega_a t)$. In general that excitation is quasiperiodic. The necessary condition for chaos is given by Eq. 8.2.5 in which $a_1 = a_2 = \cdots = a_K = 0$. We choose A_0 so that, for $A_a = 0$, the motion is incapable of crossing the potential barrier and is therefore confined to one well; in accordance with Melnikov theory this will be the case if the Melnikov inequality 8.2.3 is not satisfied. We now add the excitation $A_a \sin(\omega t)$. For a certain region R_a of the parameter space $[A_a, \omega_a]$ the system can experience chaotic motion with escapes over the potential barrier. The Melnikov scale factor $S(\omega)$ provides the information needed to select frequencies ω_a such that the added excitation will be most effective in inducing such motion. It follows from Eq. 8.2.5 (in which $a_1 = a_2 = \cdots = a_K = 0$) that, for this to be the case, ω_a should be equal or close to the frequency for which $S(\omega)$ is largest.

For chaotic motions the spectral density has (1) peaks at the fundamental excitation frequencies ω_0 and ω_a and at linear combinations thereof, and (2) a broadband portion due to the chaotic nature of the response. Given the existence of a broadband spectrum—similar to the broadband spectrum present in systems exhibiting classical stochastic resonance or in nonchaotic systems excited by broadband noise—it is reasonable to expect that the synchronization-like phenomenon noted earlier for classical stochastic resonance will similarly occur for the deterministically excited chaotic system.

This was verified by numerical simulation for a large number of cases. As a typical example, we consider the case $\beta = 0.316$, $A_0 = 0.095$, $\omega_0 = 0.0632$ (for these values Eq. 8.2.3 is not satisfied), and $\omega_a = 1.1$. Spectral densities of motions with these parameters and the parameters $A_a = 0.263$, $A_a = 0.287$, and $A_a = 0.332$, are shown in Figs. 8.1a, b, and c, respectively. Note that, owing to the presence of a broadband portion in the output spectrum, a signal-to-noise ratio can be defined for the output just as for classical stochastic resonance. For Fig. 8.1b, $\alpha = 0.0672$ is close to the signal frequency $\omega_0 = 0.0632$. The energy in the broadband portion of the spectrum is depleted, while the energy at the signal's frequency is enhanced, with respect to their counterparts in Fig. 8.1a, for which $\alpha = 0.0395$. The synchronization-like phenomenon is clearly evident in Fig. 8.1b. We also verified that the motions of Figs. 8.1a, b, and c are chaotic (i.e., their largest Lyapounov exponents, estimated as in Fig. 3.3, are positive).

Figure 8.2 shows the dependence of the signal-to-noise ratio on A_a. The plot of Fig. 8.2 is similar qualitatively to plots of signal-to-noise ratio versus noise intensity for classical stochastic resonance, as will be seen in the next section (Fig. 8.5 below).

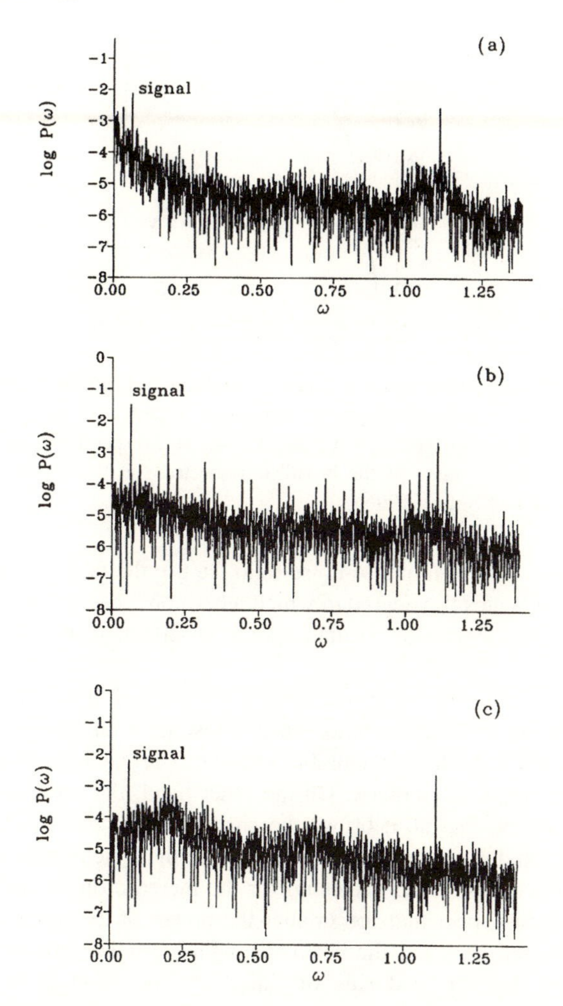

Figure 8.1. Power spectra $P(\omega)$ of system response with deterministic excitation $A_0 \sin(\omega_0 t) + A_a \sin(\omega_a t)$, $A_0 = 0.095$, $\omega_0 = 0.0632$, $\omega_a = 1.1$ (logarithms in base 10). (a) $A_a = 0.263$; (b) $A_a = 0.287$; (c) $A_a = 0.332$. The motion is chaotic in all cases. In case (b) the escape rate is approximately equal to the signal frequency. A synchronization-like phenomenon occurs that transfers broadband energy associated with the chaotic motion to the signal frequency (after Franaszek and Simiu, 1996a).

8.4 NOISE SPECTRUM EFFECT ON SIGNAL-TO-NOISE RATIO FOR CLASSICAL STOCHASTIC RESONANCE

We now consider a system excited by noise and a harmonic signal which, by itself, cannot induce escapes. To assess the effect of the shape of the

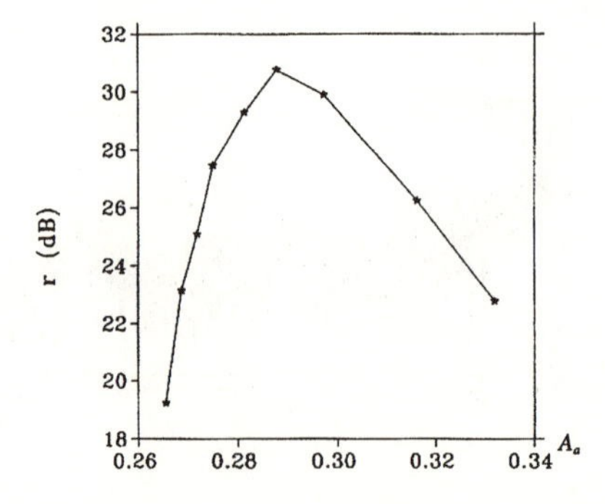

Figure 8.2. Signal-to-noise ratio r versus amplitude of added harmonic excitation A_a for system with deterministic excitation $A_0 \sin(\omega_0 t) + A_a \sin(\omega_a t)$. $r = 10 \log_{10}[P(\omega_0)/P_n(\omega_0)]$, where $P(\omega_0)$ and $P_n(\omega_0)$ are, respectively, the spectrum of the output and the ordinate of the broadband spectrum at the signal frequency ω_0 (after Franaszek and Simiu, 1996a).

noise spectrum on the signal-to-noise ratio we use the fact that the Melnikov scale factor indicates the degree to which a harmonic excitation, or frequency components of a stochastic excitation, can be effective in inducing chaotic behavior.

On the one hand the increase of the noise excitation has an unfavorable effect on the signal-to-noise ratio, since it results in an increase of the output noise level. It is this unfavorable effect that renders classical stochastic resonance an apparent paradox. On the other hand, the noise excitation has a favorable effect, that is, it brings the characteristic rate α in line with the frequency ω_0 and thus allows the occurrence of the synchronization-like phenomenon which more than makes up for the increase of the output noise. It is reasonable to expect that the smaller the power of the noise that helps to bring about a rate $\alpha \approx \omega_0$, the better the signal-to-noise ratio will be.

The larger the left-hand side of Eq. 8.2.5, the stronger is the chaotic transport across the pseudoseparatrix, and therefore the larger is the rate α. Recall that in Eq. 8.2.5 (in which it is now assumed that $A_a = 0$), $a_k = \gamma(2\Psi_0(\omega)\Delta\omega/(2\pi))^{1/2}$. Therefore, for any given power γ^2 of the stochastic excitation, the left-hand side of Eq. 8.2.5 and the rate α increase as the integral

$$I = \int_0^{\omega_{\text{cut}}} \Psi_0(\omega) S^2(\omega)\, d\omega \tag{8.4.1}$$

increases (note that in Eq. 8.4.1 $S(\omega)$ is defined as in Eq. 8.2.3). The integrand in Eq. 8.4.1 is proportional to the contribution of the stochastic excitation to the spectral ordinate of the Melnikov process at frequency ω. The frequency ω_{cut} is the maximum or cutoff frequency, beyond which the spectral density of the excitation is negligible or vanishes; see also Eq. 5.6.6.

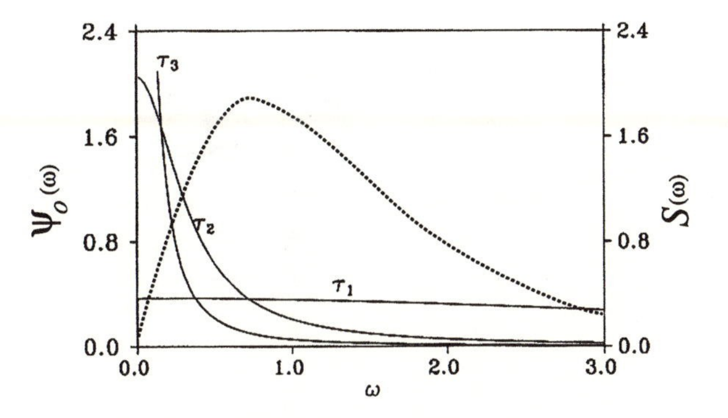

Figure 8.3. Melnikov scale factor $S(\omega)$ for $\gamma = 1$ (dotted line) and normalized power spectra $\Psi_0(\omega)$ of stochastic excitation process $G(t)$ for three values of parameter τ in Eq. 4.6.9 (solid lines): $\tau_1 = 0.2$, $\tau_2 = 3.0$, $\tau_3 = 12.0$.

Equation 8.4.1 yields the interesting qualitative result that, for a given Melnikov scale factor and fixed power of the stochastic excitation, the rate α increases as the power spectrum of the excitation is distributed nearer the frequency of the Melnikov scale factor's peak ω_{pk}, the greatest effectiveness being achieved by a single component with frequency equal to ω_{pk}.

We now illustrate the usefulness of this result for a system with classical stochastic resonance (i.e., one for which in Eq. 8.2.6 $A_a = 0, \gamma \neq 0$). We assume $G(t)$ has the spectral distribution

$$\Psi_0(\omega) = \mu/(1 + \tau^2\omega^2) \tag{8.4.2}$$

cut off at the frequency ω_{cut}; μ is a normalization factor such that the variance of $G(t)$ is unity. Figure 8.3 shows spectra $\Psi_0(\omega)$ for three values of τ, the cutoff frequency being $\omega_{cut} = 3.0$. Also shown in Fig. 8.3 is a plot of the Melnikov scale factor $S(\omega)$ (Eq. 2.4.23) defined as in Eq. 8.2.3. Provided that the cutoff frequency is sufficiently high, e.g., $\omega_{cut} \gtrsim 3.0$, its value does not significantly affect the results since for $\omega > \omega_{cut}$ $S(\omega)$ is small. We are interested in the effect of the parameter τ (i.e., of the shape of the noise spectrum) on the peak signal-to-noise ratio.

We examine first the case $\tau = \tau_1 = 0.2$. Examples of averaged output spectra $P(\omega)$ for $A_0 = 0.3, \omega_0 = 0.069, \omega_{cut} = 3.0$, and $\beta = 0.25$ are shown in Figs. 8.4a, b, and c for power $\gamma^2 = 0.005, \gamma^2 = 0.02$, and $\gamma^2 = 0.11$, respectively. The averaging was performed over 225 noise realizations approximated by Eq. 8.2.7 with $100 < K < 500$. Note that $A_0 < 4\beta/[3S(\omega_0)|_{\gamma=1}]$, so that no chaotic behavior can be induced by the periodic signal alone. However, it was verified that, for the noise realizations used to obtain the results of Figs. 8.4a,b,c, the Melnikov inequality (Eq. 8.2.5) was satisfied, and the respective motions were chaotic. Energy transfer to the

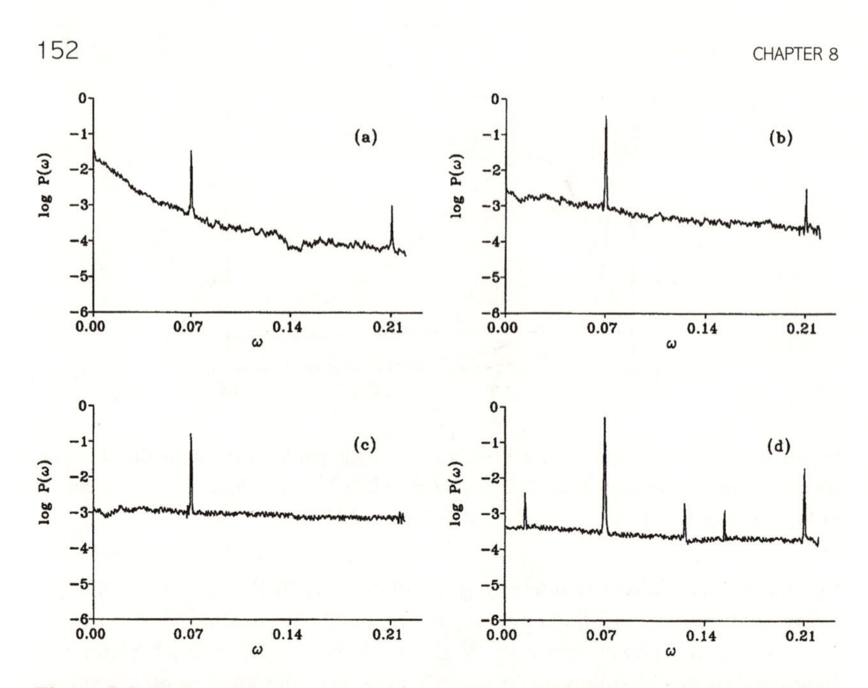

Figure 8.4. Averaged power spectra of output for stochastically excited system: (a)–(c) increasing noise strength γ and amplitude of added harmonic $A_a = 0$; (d) same noise strength γ as in (a), and $A_a = 0.23$. Parameter $\tau = 0.2$ in all cases (logarithms in base 10). Note that the signal-to-noise is increased more effectively by the added harmonic excitation (case d) than by the optimum increase of the noise intensity (case b).

signal frequency was found to be highest for the excitation inducing a rate α close to the signal frequency. This occurred for the case represented in Fig. 8.4b.

The dependence of signal-to-noise ratio on noise intensity is plotted in Fig. 8.5. Figure 8.5 also shows similar plots for $\tau = 3$ and $\tau = 12$, the parameters A_0, ω_0, ω_{cut}, and β being the same as for the case $\tau = 0.2$. For $\tau = 0.2$, $\tau = 3$, and $\tau = 12$, $I = 0.626$, $I = 0.411$, and $I = 0.157$, respectively. As expected, the peak signal-to-noise ratio is smaller and occurs at higher values of γ for the larger values of τ, that is, for spectral shapes to which there correspond smaller values of the integral I.

8.5 SYSTEM WITH HARMONIC SIGNAL AND NOISE: SIGNAL-TO-NOISE RATIO ENHANCEMENT THROUGH THE ADDITION OF A HARMONIC EXCITATION

The results of the preceding sections suggest the following method for increasing the signal-to-noise ratio. Assume that in Eq. 8.2.6 $A_a = 0$, and

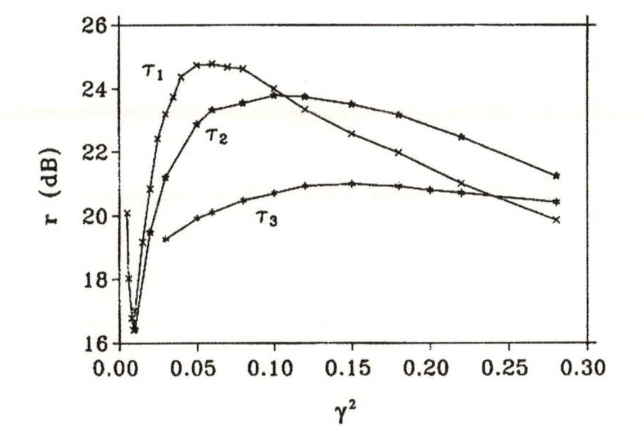

Figure 8.5. Signal-to-noise ratio r versus noise power γ^2 for three values of τ: $\tau_1 = 0.2$, $\tau_2 = 3.0$, $\tau_3 = 12.0$.

that for a set of values A_0, ω_0, β, and γ the system has low signal-to-noise ratio. We could increase the signal-to-noise ratio by increasing γ, as illustrated in the preceding section. However, it is more effective to do so by keeping γ unchanged and adding an excitation $A_a \sin(\omega_a t)$ such that (1) ω_a is equal or close to the frequency of $S(\omega)$'s peak and (2) A_a is so chosen as to bring about a characteristic rate α comparable to the signal frequency. An example is shown in Fig. 8.4d, for which all parameters and the spectrum $\Psi_0(\omega)$ are the same as for Fig. 8.4a, except that the system is subjected to an added excitation with amplitude $A_a = 0.23$ and frequency $\omega_a = 1.1$. A comparison between Figs. 8.4d and 8.4b shows that this approach to increasing the signal-to-noise ratio is superior. Note that the added harmonic excitation induces subharmonics and superharmonics that are well separated from the signal and can therefore be filtered out if necessary.

8.6 NONLINEAR TRANSDUCING DEVICE FOR ENHANCING SIGNAL-TO-NOISE RATIO

Based on the method discussed in Section 8.5 we now describe the principle of a proposed nonlinear transducing device for improving signal-to-noise ratio. Assume we have a signal with unsatisfactory signal-to-noise ratio. We use that signal and the attendant noise—from which we filter out components with frequencies exceeding, say, three times the signal frequency—as input to excite a transducing device. The signal-to-noise ratio of the device's output will in general be poor. However, under certain conditions it can be improved by adding to the input excitation just described a harmonic excitation with frequency equal or close to the frequency of the Melnikov scale factor's peak.

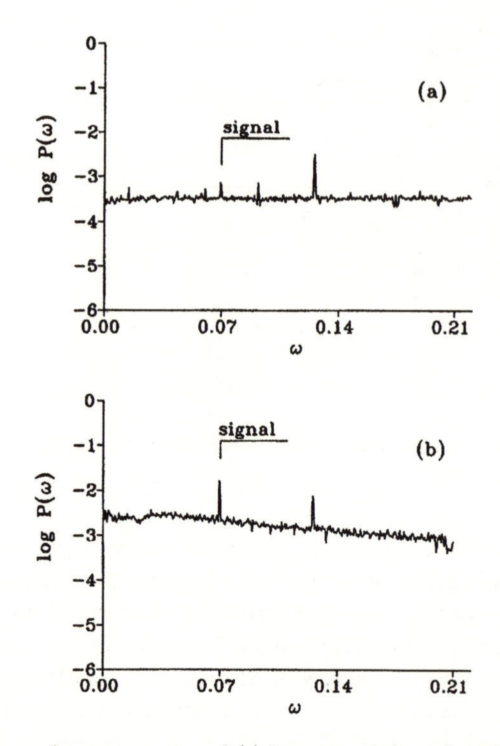

Figure 8.6. Averaged power spectra of (a) input consisting of stochastic excitation, harmonic signal with frequency $\omega_0 = 0.069$, and additional harmonic excitation with frequency $\omega_a = 1.1$; (b) output of transducing device (logarithms in base 10).

The role of the added harmonic excitation is to bring about a chaotic motion with characteristic rate α close to the signal frequency, and the consequent transfer of energy from the broadband spectrum to the signal frequency.

To illustrate the principle of the proposed device, we show in Fig. 8.6a the spectral density of a signal $A_0 \sin(\omega t)$, $A_0 = 0.05$, $\omega_0 = 0.069$, largely drowned in colored noise $\gamma G(t)$ with $\tau = 0.2$ (Eq. 8.4.2 and Fig. 8.3), $\gamma^2 = 0.36$. By using a low passband filter, we filter out the noise components with frequencies larger than three times the frequency of the signal. The signal and the noise left after the filtering (i.e., the noise $\gamma G(t)H(3\omega_0)$, where H denotes the Heaviside step function) are used as input excitation to a standard Duffing–Holmes oscillator (Eqs. 2.5.15 in which $a = b = 1$) with parameter $\beta = 0.25$. For $A_a = 0$ the signal-to-noise ratio for the output turns out not to be better than for the input. However, by adding to that input an excitation $A_a \sin(\omega_a t)$, where in our example $A_a = 0.23$ and $\omega_a = 1.1$, we obtain the spectrum with the enhanced component at the signal frequency shown in Fig. 8.6b.

8.7 CONCLUDING REMARKS

The Melnikov approach applied to the stochastic resonance problem provides a unifying framework wherein classical stochastic resonance—the enhancement of the signal-to-noise ratio achieved by increasing the noise intensity—is viewed as a particular case of a type of chaotic dynamics that includes, as another particular case, the enhancement of the signal-to-noise ratio achieved by adding a harmonic excitation while leaving the noise unchanged. Two facts are relevant to the use of Melnikov theory in the context of stochastic resonance. First, independently of whether the excitation is deterministic, stochastic, or mixed, the output motion is chaotic and therefore has a broadband spectrum. This makes possible the occurrence of a synchronization-like phenomenon that is the key to the enhancement of the signal-to-noise ratio. Second, the effectiveness in promoting chaos of a harmonic excitation with given frequency or of an added stochastic excitation with given frequency content depends on the system's Melnikov scale factor. The Melnikov approach provides the basis of an alternative mechanism for enhancing signal-to-noise ratio where the noise intensity is left unchanged, rather than being increased, and a harmonic excitation is added to the system. This alternative mechanism is more effective—allows a better signal-to-noise ratio to be obtained—than the mechanism that relies on increasing the noise intensity, and underlies the principle of a proposed device that accepts a signal with low signal-to-noise ratio and converts it into an output for which the signal-to-noise ratio is improved. Melnikov theory also provides a convenient method for assessing the effect of the spectral density of the noise on signal-to-noise enhancement in classical stochastic resonance. From Melnikov theory it follows that the enhancement of the signal-to-noise ratio is stronger as the spectral power of the noise is distributed closer to the Melnikov's scale factor's peak.

Chapter Nine

Cutoff Frequency of Experimentally Generated Noise for a First-Order Dynamical System

9.1 INTRODUCTION

For stochastic systems with rates of escape that must be determined experimentally it is necessary to generate experimental noise with a sufficiently high cutoff frequency, so that the effect on the experimental results of suppressing higher-frequency noise components is negligibly small. In this chapter we show that the Melnikov method can be used to calculate an appropriate cutoff frequency for the first-order system

$$\dot{x}(t) = -V'(x) + \xi(t), \qquad (9.1.1)$$

where $V(x)$ is a multiwell potential, $\xi(t)$ is a process satisfying the equation written formally as

$$\dot{\xi}(t) = -\alpha\xi(t) + \alpha\sqrt{D}w(t), \qquad t \geq 0, \quad \xi(0) = 0, \qquad (9.1.2)$$

$\alpha > 0, D > 0$, and $w(t)$ denotes white Gaussian noise with autocovariance $\delta(t)$. It can be shown that the process $\xi(t)$ has exponentially decaying correlation

$$C(s) = (\alpha D/2)\exp(-\alpha|s|) \qquad (9.1.3)$$

(Larson and Shubert, 1979, vol. 2, p. 403, and *Encyclopedia of Statistical Sciences*, 1985, vol. 6, p. 518), and one-sided spectral density

$$\Psi_0(\omega) = 2D/(1 + \omega^2/\alpha^2), \qquad (9.1.4)$$

that is, $\xi(t)$ is an Ornstein-Uhlenbeck process (see Section 4.2.3.2). In Section 9.2 it is shown that Eq. 9.1.1 in which the noise correlation is defined by Eq. 9.1.3 can be transformed into a second-order system possessing a Melnikov process. The corresponding expression for the Melnikov scale factor can be used to select an appropriate cutoff frequency for the experimentally generated noise.

9.2 TRANSFORMED EQUATION EXCITED BY WHITE NOISE

In general, because Eq. 9.1.1 is of first order, it cannot exhibit expansion, contraction, and folding patterns associated with chaotic motions in planar or higher-dimensional systems, and it does not have a Melnikov process. However, in the particular case where the colored noise excitation $\xi(t)$ is an Ornstein-Uhlenbeck process, Eq. 9.1.1 can be transformed into a second-order dynamical system that has a Melnikov process (Franaszek, 1996).

To show this we use the notation $y(t) = V'(x) + \xi(t)$, and rewrite Eqs. 9.1.1 and 9.1.2 in the form

$$\dot{x}(t) = y(t), \tag{9.2.1a}$$

$$\dot{y}(t) = -[V''(x) + \alpha]y(t) - \alpha V'(x) + \alpha D^{1/2} w(t). \tag{9.2.1b}$$

Equations 9.2.1 describe a system with potential $\alpha V(x)$, nonlinear damping $V''(x) + \alpha$, and excitation $\alpha D^{1/2} w(t)$. If $\alpha V(x)$ is a multiwell potential, Eqs. 9.2.1 possess a Melnikov process.

In the following we consider the standard Duffing–Holmes oscillator, that is, we assume $V(x) = -x^2/2 + x^4/4$. For Eqs. 9.2.1 we denote the Melnikov scale factor corresponding to a harmonic excitation with unit amplitude by $S_M(\omega)$. We have

$$S_M(\omega) = (2/\alpha)^{1/2} \pi\omega \operatorname{sech}\big(\pi\omega/(2\alpha^{1/2})\big) \tag{9.2.2}$$

(Eq. 2.5.17 in which $a = b = \alpha$ and $\gamma = 1$). Figure 9.1 represents $S_M(\omega)$ for three values of the correlation time parameter α. From Eq. 9.2.2 it follows that the frequency for which $S_M(\omega)$ reaches its maximum is $\omega_{\max} = c\alpha^{1/2}$, where $c \approx 0.76$. This result is obtained by equating to zero the derivative with respect to ω of Eq. 9.2.2. The contribution of white noise components with frequencies larger than the cutoff frequency ω_{cut} is negligible if for those frequencies the Melnikov scale factor $S_M(\omega)$ is negligibly small. The cutoff frequency should therefore satisfy the inequality

$$\omega_{\text{cut}} \geq K(\alpha)\alpha^{1/2} \tag{9.2.3}$$

where $K(\alpha)$ is sufficiently large.

This was verified by numerical simulations of Eqs. 9.2.1 (Franaszek, 1996), where the white noise $w(t)$ was approximated as shown in Section 4.2.3.2. The simulation parameters were $\alpha = 1.6 \times 10^3$, $N = 300$ and $D = 0.495$. The ratio between the mean escape rate obtained by simulation r (based on 100 realizations of the noise) and a theoretical mean escape rate r_t (Lindenberg, West, and Masoliver, 1989) is shown in Fig. 9.2 as a function

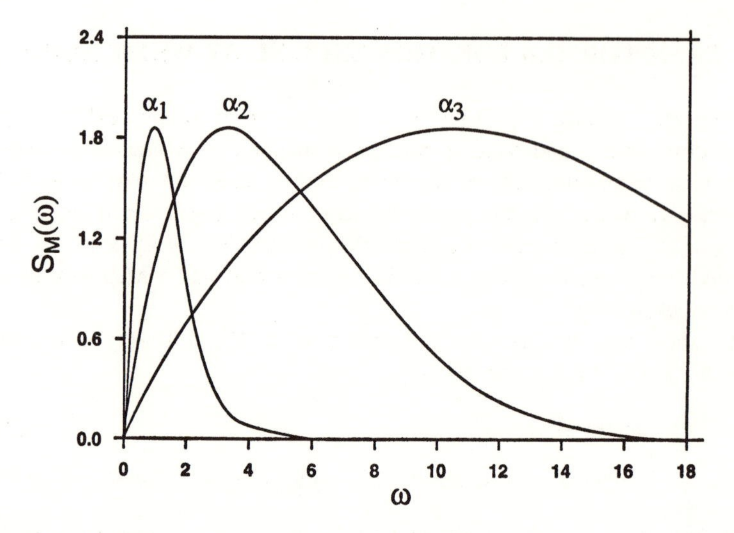

Figure 9.1. Melnikov scale factor $S_M(\omega)$ for system 9.2.1 and three values of colored noise parameter, $\alpha_1 = 3.4$, $\alpha_2 = 19$, $\alpha_3 = 190$; see Eq. 9.2.2.

of K. Figure 9.2 shows that a cutoff frequency for which $K > 2.5$ is acceptable, that is, for such a cutoff frequency the experimental value r is sufficiently close to the theoretical value r_t.

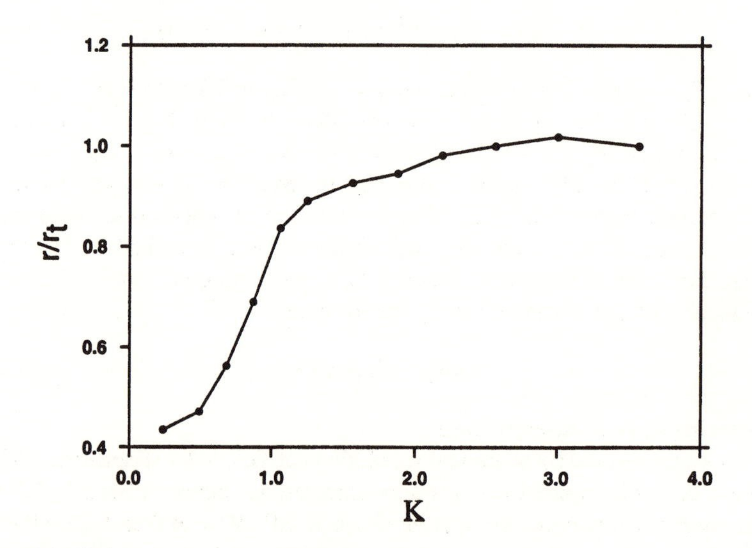

Figure 9.2. Mean escape rate r obtained by numerical simulations for various values of parameter K. The cutoff frequency is $\omega_{\text{cut}} = K\alpha^{1/2}$, $\alpha = 1.6 \times 10^3$; r_t is the theoretical mean escape rate (after Franaszek, 1996).

Chapter Ten

Snap-Through of Transversely Excited Buckled Column

This section presents a structural/mechanical engineering application of the Melnikov method. We seek to obtain criteria for the occurrence of stochastically induced transitions in a spatially extended dynamical system (i.e., a system governed by a partial differential equation with space and time coordinates). The system we consider is a buckled column with continuous mass, subjected to a transverse, continuously distributed force that varies randomly with time. The force may be due, for example, to seismic motion, pressures induced by air flow turbulence, or effects arising in hydrodynamical systems.

In the absence of axial and transverse loading the axis of the column is assumed to be straight. As a result of buckling induced by a constant axial force the column is deformed—bent—on one of the two sides of its original, undeformed axis. The time-dependent transverse loading further contributes to the deformation. As long as the transverse loading is sufficiently small the column's deformations occur only on one side of the undeformed axis. However, if the transverse loading is sufficiently large, snap-through occurs, that is, the motion undergoes transitions similar to those of Fig. 1.2c. Motion occurring solely on one side of the undeformed axis can be viewed as motion within a well of a double-well potential, and snap-through can be viewed as motion across the potential barrier separating the two wells.

For a deterministic counterpart of our problem—a continuous buckled column subjected to a transverse, uniformly distributed load varying harmonically in time—a Melnikov-based necessary condition for the occurrence of snap-through was obtained by Holmes and Marsden (1981) under the simplifying assumption that deformations are small. This condition is a used as a building block for the extension of the Melnikov method to the case of random transverse loading.

We first present the equation of motion of the column and briefly review the Melnikov approach developed by Holmes and Marsden (1981) for the harmonic excitation case. We then show that the Melnikov approach can be extended to the case of nonharmonic excitation, including random excitation. We consider a continuous buckled column excited by (a) colored Gaussian noise and (b) dichotomous noise. For case (a), the probability that snap-through can occur during any specified time interval is always larger than zero, and if escapes are rare events an upper bound for that probability can be obtained in closed form. For case (b), results must be obtained numerically; in addition, the Melnikov necessary condition for chaos yields a simple criterion that guarantees the nonoccurrence of snap-through. We conclude the chapter with a numerical example.

10.1 EQUATION OF MOTION

Assume that (a) the mechanical properties of the column are uniform over its length, (b) the behavior of the material is linearly elastic, (c) the column, which was originally straight, has been compressed past the critical buckling load P_{cr} to a static deflected position, (d) the column deformations are sufficiently small that, in the Taylor expansion of the projection of the elemental deformed column length on the line joining the column supports, terms of power higher than 2 can be neglected, and (e) the column is hinged at both ends. The nondimensionalized equation of the column is (Tseng and Dugundji, 1971; Holmes and Marsden, 1981)

$$z_{tt} + z_{yyyy} + \left\{\Gamma - \xi \int_0^1 z_y^2(\zeta, t)\, d\zeta\right\} z_{yy} = \epsilon\{R(y, t) - \beta z_t\}, \quad (10.1.1a)$$

$$R(y, t) = \gamma(y)\cos(\omega_0 t) + \rho(y)G(t), \quad (10.1.1b)$$

where $z(y, t) = Z(Y, \tau)/\Delta$ is the dimensionless deflection, Z is the deflection at time τ, Y is the coordinate along the column length ℓ, $y = Y/\ell$, $\Delta = Z_0(\ell/2)$ is the static deflection of the column $Z_0(Y)$ at coordinate $Y = \ell/2$,

$$\Gamma = P_0\ell^2/EI, \quad (10.1.1c)$$

E is Young's modulus, I is the moment of inertia of the column cross-section,

$$P_0 = P_{cr} + [EA/2\ell]\int_0^\ell (dZ_o/dY)^2 dY, \quad (10.1.1d)$$

$$P_{cr} = \pi^2 EI/\ell^2 \quad (10.1.1e)$$

is the Euler critical buckling load, A is the column's cross-sectional area, $\xi = ½\Delta^2 A/I$, $\epsilon\beta = c\ell^2/[mEI]^{1/2}$, c is the viscous damping coefficient, m is the column mass per unit length, $t = \omega_1\tau$ is the nondimensional time, τ is the dimensional time, $\omega_1^2 = (EI/\ell^4 m)$, $\epsilon\gamma(y) = f(Y)\ell^4/(EI\Delta)$, $f(Y)$ is the amplitude of the harmonic force per unit length, $G(t)$ is a nondimensional nonperiodic function, $\epsilon\rho(y) = s(Y)\ell^4/(EI\Delta)$, and $s(Y)$ is a measure of the nonperiodic force per unit length. The assumption that the ends of the column are hinged and the fact that for small deformations bending moments are proportional to z_{yy} together imply the boundary conditions $z(0,t) = z(1,t) = z_{yy}(0,t) = z_{yy}(1,t) = 0$. The initial deflected shape of the column $Z(Y,0)$ is assumed to be

$$Z_0(Y) = \Delta \sin(\pi Y/\ell). \tag{10.1.1f}$$

Recall that the distance between the ends of the columns is fixed after the initial static compression. The second term in the right-hand side of Eq. 10.1.1d and the nonlinear term in Eq. 10.1.1a are due to the compression of the column past the critical buckling load, which results in an initial static deflection (10.1.1f).

The static, linearized, and unperturbed counterpart of Eq. 10.1.1a for the case $\Delta = 0$ is

$$z_{yyyy} + \pi^2 z_{yy} = 0, \tag{10.1.2}$$

where, by virtue of Eqs. 10.1.1c–f and of the fact that $\Delta = 0$, we have $\Gamma = \pi^2$. It is a classic result associated with Euler that, in addition to the trivial solution $z = 0$, Eq. 10.1.2 has nontrivial solutions satisfying the boundary conditions, the eigenfunctions $\sin(j\pi y)$ $(j = 1, 2, \ldots)$.

The functions $\gamma(y)$ and $\rho(y)$ can be expanded in the Fourier series

$$\gamma(y) = \gamma_0 + \sum_{n=1}^{\infty} \{\alpha_{\gamma_n} \sin(n\pi y) + \beta_{\gamma_n} \cos(n\pi y)\}, \tag{10.1.3a}$$

$$\rho(y) = \rho_0 + \sum_{n=1}^{\infty} \{\alpha_{\rho_n} \sin(n\pi y) + \beta_{\rho_n} \cos(n\pi y)\}. \tag{10.1.3b}$$

10.2 HARMONIC FORCING

The case $\rho \equiv 0$, $\gamma \neq 0$ (Eq. 10.1.1b) was studied by Holmes and Marsden (1981). In this section we briefly summarize their results.

The function $z(y,t)$ is expanded in the eigenfunctions of the linearized system as follows:

$$z(y,t) = \sum_{j=1}^{\infty} a_j(t) \sin(j\pi y). \tag{10.2.1}$$

Substitution of Eq. 10.2.1 into Eq. 10.1.1a, and the application of the Galerkin method, yield

$$\ddot{a}_j + \epsilon\beta\dot{a}_j + (j\pi)^2\left\{(j\pi)^2 - \left[\Gamma - (\xi\pi^2/2)\sum_{k=1,2,\dots} k^2 a_k^2\right]\right\}a_j$$
$$= 2\epsilon\phi_j\cos(\omega_0 t), \tag{10.2.2}$$

where $\phi_j = \int_0^1 \gamma(y)\sin(j\pi y)\,dy$.

The unperturbed counterpart of Eq. 10.2.2 has a fixed point at the origin $\{a_1 = 0, \dot{a}_1 = 0, a_2 = 0, \dot{a}_2 = 0, \dots\}$. The corresponding first variational equation is

$$\delta\ddot{a}_j + (j\pi)^2[(j\pi)^2 - \Gamma]\delta a_j = 0. \tag{10.2.3a}$$

Its eigenvalues satisfy the equations

$$\lambda_j^2 + (j\pi)^4 - \Gamma(j\pi)^2 = 0,\ j = 1, 2, \dots,\ \text{or} \tag{10.2.3b}$$

$$\lambda_{j1,2} = \pm j\pi\left(\Gamma - (j\pi)^2\right)^{1/2}. \tag{10.2.3c}$$

From Eqs. 10.1.1c–e it follows immediately that

$$\Gamma = \pi^2 + \pi^2\xi/2 > \pi^2. \tag{10.2.3d}$$

In addition, we assume that the deflections, and therefore ξ, are sufficiently small that

$$\pi^2 < \Gamma < 4\pi^2. \tag{10.2.4}$$

The results that follow depend on this assumption. Equations 10.2.3c and 10.2.4 imply that the eigenvalues $\lambda_{j1,2}$ are real and of opposite signs for $j = 1$ and are purely imaginary for all $j \geq 2$. It is shown by Holmes and Marsden (1981, pp. 150, 151) that this in effect reduces the Melnikov problem to a planar one, and implies that (a) the Melnikov function for Eq. 10.1.1a is

$$M(t) = \int_{-\infty}^{\infty}\int_0^1 \left[R(y,\zeta)\dot{z}_h(y, t-\zeta) - \beta\dot{z}_h^2(y, t-\zeta)\right]dy\,d\zeta \tag{10.2.5}$$

where $R(y, t)$ is given by Eq. 10.1.1b in which $\rho(y) \equiv 0$; (b) the homoclinic orbit of the unperturbed system has coordinates

$$z_h(y, t) = \pm\sqrt{2}\sin(\pi y)\operatorname{sech}\left(t\pi^2\xi^{1/2}/\sqrt{2}\right), \tag{10.2.6a}$$

$$\dot{z}_h(y, t) = \mp\pi^2\xi^{1/2}\sin(\pi y)\operatorname{sech}\left(t\pi^2\xi^{1/2}/\sqrt{2}\right)$$
$$\times\tanh\left(t\pi^2\xi^{1/2}/\sqrt{2}\right); \tag{10.2.6b}$$

and (c) the expression for the Melnikov function is

$$M(t_0) = k_1\beta + (\alpha_{\gamma_1} + 4\gamma_0/\pi)k_2(\omega_0 t_0), \tag{10.2.7a}$$

$$k_1 = -\sqrt{2}\pi^2\xi^{1/2}/3,$$

$$k_2(\omega_0) = -(1/\sqrt{2})\omega_0 \operatorname{sech}\big[\omega_0/\big(\sqrt{2}\pi\xi^{1/2}\big)\big]. \tag{10.2.7b, c}$$

For sufficiently small ϵ, the stable and unstable manifolds of the perturbed system intersect transversely if $M(t_0)$ has simple zeros. The dynamics of the system then contains a horseshoe, which is associated with the possible existence of a chaotic attractor.

10.3 STOCHASTIC FORCING. NONRESONANCE CONDITIONS. MELNIKOV PROCESSESES FOR GAUSSIAN AND DICHOTOMOUS NOISE

If the nonresonance condition $\omega_0^2 \neq \lambda_j^2$ holds Eq. 10.2.2 has unique solutions of $O(\epsilon)$. Otherwise Eq. 10.2.2 has solutions of $O(1)$. This would violate a basic assumption of Melnikov theory. If the excitation is quasiperiodic the nonresonance condition must be satisfied for each component of the quasiperiodic sum.

If the forcing is a stochastic process, it has components over a continuous range of frequencies, that is, it may also have elemental components with frequencies equal to the natural frequencies of the linearized counterpart of Eq. 10.2.2. In this case, for white noise, it can be shown that the solutions of Eq. 10.2.2 are $O(\epsilon^{1/2})$ (see, e.g., Meirovich, 1967, p. 501). For sufficiently small ϵ the solutions will therefore be as small as desired, and nonresonance conditions are not required for Melnikov theory to be applicable. However, if in Eq. 10.1.1b $\gamma(y) \neq 0$, nonresonance conditions must be satisfied for the harmonic forcing with frequency ω_0. A similar statement holds for quasiperiodic excitation.

For the system excited by a stochastic process the Melnikov process has the same form as Eq. 10.2.5, in which $R(y, t)$ is given by Eq. 10.1.1b. If $\gamma(y) \equiv 0$, and $G(t)$ is Gaussian with zero mean, unit variance, and spectral density $\Psi_0(\omega)$, the Melnikov process is Gaussian with spectral density, expectation, and variance

$$\Psi_M(\omega) = (\alpha_{\rho_1} + 4\rho_0/\pi)^2 k_2^2(\omega)\Psi_0(\omega), \qquad E[M] = k_1\beta, \tag{10.3.1a,b}$$

$$\operatorname{Var}[M] = (1/2\pi)\int_0^\infty \Psi_M(\omega)\, d\omega. \tag{10.3.1c}$$

Equations 10.3.1 may be used to estimate lower bounds τ_M for the mean time τ_t between snap-through events (Eqs. 5.5.2, 5.5.3), and lower bounds for the probability that no snap-through occurs during a time interval T (Eq. 5.5.4).

If the noise is of the coin-toss, square-wave dichotomous type (see Eq. 4.2.16 and Fig. 4.5) with amplitude $\rho(y)$ defined as in Eq. 10.1.3b, the Melnikov process can be written as

$$M(t_0) = k_1\beta + \sqrt{2}(\alpha_{\rho_1} + 4\rho_0/\pi)F(t\pi^2\xi^{1/2}), \tag{10.3.2}$$

where $F(t)$ is defined by Eq. 5.4.6. This follows from Eq. 10.2.5 in which $R(y, \zeta)$ denotes the coin-toss dichotomous noise, Eq. 4.2.16 (or Eq. 5.4.1), and Eqs. 10.2.6b. The area under the curve $h(\zeta)$ in a half-plane is $\lim_{t\to\infty} z_h(y = 1/2, t) - z_h(y = 1/2, 0) = \sqrt{2}$. Since the largest value of $F(t\pi^2\xi^{1/2})$ occurs when $\alpha = 0$, $a_n = 1$ for all n such that $t > 0$, and $a_n = -1$ for all n such that $t < 0$, we have

$$-2\sqrt{2} < F\left[t\pi^2\xi^{1/2}\right] < 2\sqrt{2}. \tag{10.3.3}$$

If $\rho(y) \equiv \rho_0$, from Eqs. 10.3.2 and 10.3.3 and the necessary condition for the occurrence of chaos we obtain the result that snap-through cannot occur if

$$\rho_0 < \rho_{0,\,\mathrm{crit}} = \pi^3\xi^{1/2}\beta/(24\sqrt{2}). \tag{10.3.4}$$

Equation 10.3.4 is consistent with the intuitively obvious fact that the larger the parameter ξ, that is, the larger the initial deflection Δ of the column, the larger is the strength ρ_0 of the dichotomous noise required to bring about column snap-through.

10.4 NUMERICAL EXAMPLE

We consider Eqs. 10.1.1, $\gamma(y) \equiv 0$, $\rho(y) \equiv \rho_0$, $\ell = 0.45$ m, and assume that the column cross section is rectangular with dimensions $h = 0.0005$ m, $b = 0.0125$ m; $E = 200{,}000$ MPa, $\Delta = 0.0005$ m, $\epsilon = 0.1$, $\beta = 0.1866$, and the excitation is dichotomous (Eq. 4.2.16). We have $A = 6.25 \times 10^{-6}$ m^2, $I = 1.30208 \times 10^{-13}$ m^4, $m = 0.04875$ kg/m, $\xi = 6.0$. From Eq. 10.3.4, $\rho_{0,\,\mathrm{crit}} = 0.41755$.

The equations of motion of the column were solved numerically for various realizations of dichotomous noise and values of the excitation amplitude

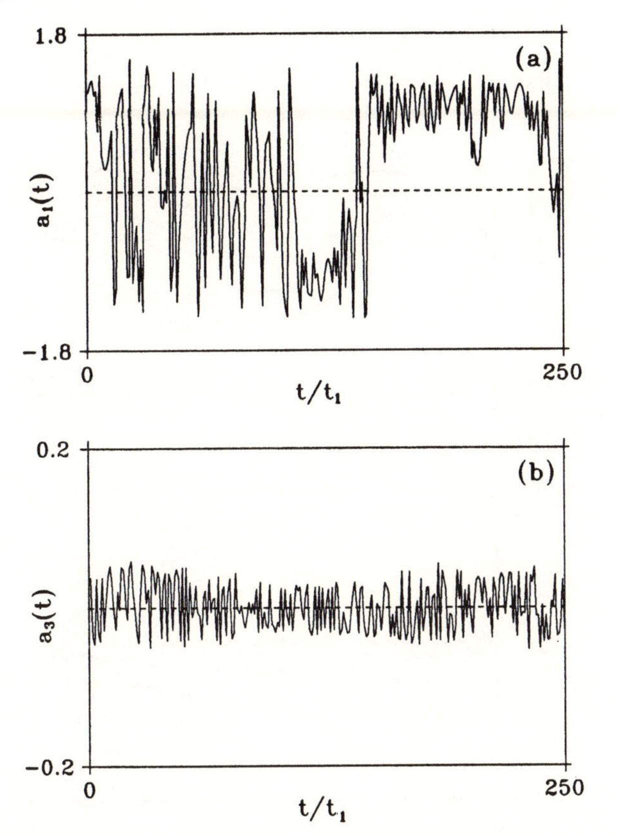

Figure 10.1. Example of steady-state time histories $a_1(t)$ and $a_3(t)$, dichotomous excitation (from Franaszek and Simiu, 1996b).

ρ_0. For the parameters just listed and $t_1 = 1.0$ (Eq. 4.2.16), the smallest excitation for which snap-through was observed was $\rho_{0,\min} = 11.4 > 0.441755$. Note that $\rho_{0,\min}$ depends on t_1. For example, for $t_1 = 0.2$, all other parameters being unchanged, $\rho_{0,\min} = 14$, that is, as expected, the dichotomous noise is less effective in inducing snap-through if $t_1 = 0.2$ than if $t_1 = 1.0$ (see Fig. 5.7 for a similar comparison).

For the parameters just listed and $t_1 = 1.0$, a steady-state time history of the amplitudes $a_1(t)$ and $a_3(t)$ for the first and third Galerkin modes, respectively, is shown in Fig. 10.1. Figure 10.2 shows the evolution in time of the column shape $z(y, t'_n) = z_1(y, t'_n) + z_3(y, t'_n)$ and the components $z_1(y, t'_n)$, $z_3(y, t'_n)$ at snap-through for six successive times $t'_n = nt_1/300$. Even-numbered components vanish, and odd-numbered components of order 5 and higher are negligible.

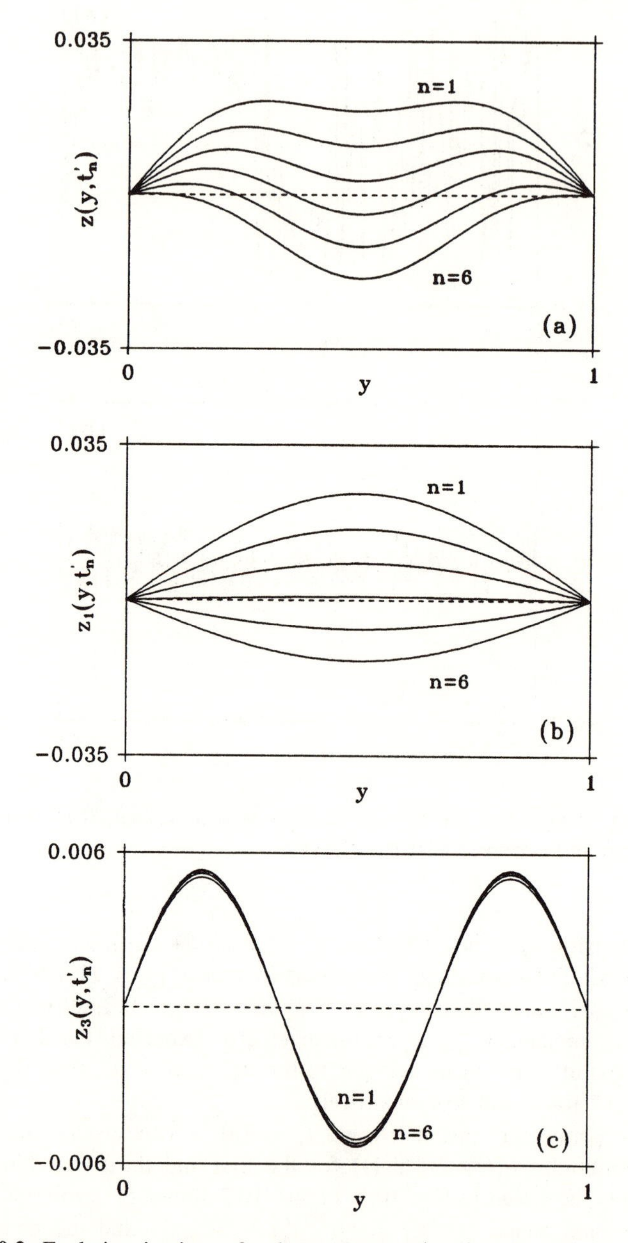

Figure 10.2. Evolution in time of column shapes $z(y, t'_n)$ and $z_j(t)$ $(j = 1, 3)$ at snap-through $(t'_n = nt_1/300)$ (from Franaszek and Simiu, 1996b).

Chapter Eleven

Wind-Induced Along-Shore Currents over a Corrugated Ocean Floor

This chapter presents an oceanographic application of the Melnikov method. We consider a simple model of mesoscale (20–500 km scale) wind-induced along-shore ocean flow over a continental margin. The ocean bottom has variable topography that slopes linearly offshore (i.e., in the direction normal to the shoreline), and along-shore sinusoidal corrugations whose amplitude vanishes at the shoreline (Fig. 11.1). The model was developed and analyzed by Allen et al. (1991) for the case of forcing by harmonically fluctuating stresses induced at the water's surface by along-shore fluctuating wind. It was extended for the case of stochastic forcing by Simiu (1996).

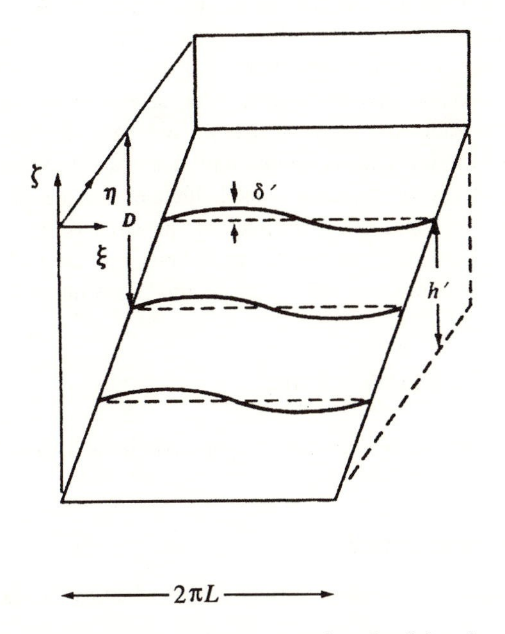

Figure 11.1. Model geometry showing one wavelength of the along-shore periodic corrugation (after Allen, Samelson, and Newberger, 1991).

Models of geophysical fluid dynamic processes have involved predominantly transient wave and vortex disturbances superposed on and interacting with large-scale mean flows. In these models the disturbances are assumed to arise as smaller-scale instabilities of the mean flows. An alternative, more recent approach applicable to some types of geophysical flow is to model them as dynamical systems that may exhibit motions with transitions (Charney and DeVore, 1979).

The model analyzed by Allen et al. (1991) is based on this approach. It was motivated by observations yielded by moored current meters, according to which mesoscale coastal velocity fields appear to contain energy in a continuous range of relatively low frequencies. Such a continuous spectral distribution may be due in part to the stochastic nature of the forcing. However, following contributions on the chaotic nature of motions in certain simplifed fluid dynamical models by, among others, Lorenz (1963), it was hypothesized that at least part of the broadband energy content of the motion is associated with chaotic motion. Allen et al. (1991) tested this hypothesis by analyzing a simplified model of the along-shore flow induced by harmonically fluctuating wind over corrugated bottom topography. It is reasonable to argue that this simplified model—to which the Melnikov method applies—captures the main features of the flow including, in particular, its chaotic or nonchaotic behavior.

In this chapter we review the model developed by Allen et al. (1991) and its extension for the case—consistent with the turbulent nature of wind fluctuations—of stochastic wind forcing (Simiu, 1996). We apply the Melnikov method and show that, like deterministic forcing, stochastic forcing is capable of inducing chaotic behavior. Results based on the assumptions of forcing with harmonic and stochastic fluctuations show that, for the same energy of the wind speed fluctuations, the propensity of the model for chaotic behavior is stronger in the stochastic case. Finally, for stochastic excitation, we provide lower bounds for the probability that no chaotic transitions will occur during a specified time interval.

11.1 OFFSHORE FLOW MODEL

The model is defined by geometrical, meteorological, and fluid dynamic assumptions.

GEOMETRICAL ASSUMPTIONS

- The corrugated ocean floor surface is described by the expression

$$h(\xi, \eta) = \eta(dh'/d\eta) + \delta' \cos(\xi/L) \qquad (11.1.1)$$

where h' is the vertical dimension, δ' the amplitude of the corrugations, L the length scale, and ξ, η, ζ are dimensional coordinates (Fig. 11.1).
- δ' varies slowly with η, and L is large in relation to the scale of the variation of δ'.

METEOROLOGICAL ASSUMPTIONS

- The wind flow is uniform in space and has mean velocity and velocity fluctuations parallel to the along-shore direction. Additional meteorological assumptions are discussed in Section 11.2.

FLUID DYNAMIC ASSUMPTIONS

- The ocean flow is driven by surface stresses due to friction at the interface between the atmosphere and the ocean. The stresses have a mean and a fluctuating component, denoted by $\epsilon\tau_0'$ and $\epsilon\tau'(t)$, respectively.
- Owing to the no-slip condition, the ocean flow has zero velocity at the ocean floor.
- The shear stresses in the ocean flow are induced by a constant eddy viscosity. (The boundary layer associated with a constant eddy viscosity is known as an Ekman layer.[1])
- The flow experiences Coriolis forces induced by the Earth's rotation. (Coriolis forces explain, for example, the vortexlike nature of cyclonic flows such as hurricanes).
- The effect of the bottom corrugations on the flow is given by a nondimensional stream function ϕ, whose truncated Fourier expansion is

$$\phi = \sqrt{2}(\phi_1 \cos(\xi/L) + \phi_2 \sin(\xi/L)) \tag{11.1.2}$$

where ϕ_1 is in phase and ϕ_2 is out of phase with the topography 11.1.1.

It is shown in Allen et al. (1991) and references quoted therein that the equations of the wind-induced ocean flow, obtained from the along-shore equation of balance of momenta averaged over a corrugation wavelength, can be written in the nondimensional form

$$\begin{aligned} \dot{x} &= \frac{\partial}{\partial y} H(x, y, z) + \epsilon g_1(x, t), \\ \dot{y} &= -\frac{\partial}{\partial x} H(x, y, z) + \epsilon g_2(y), \\ \dot{z} &= \epsilon g_3(x, z, t), \end{aligned} \tag{11.1.3}$$

[1] See, e.g., Simiu and Scanlan (1996), p. 39.

where the overdot denotes differentiation with respect to t, $t = t'U_0/L$, t' is the dimensional time, x is the nondimensionalized along-shore ocean flow velocity (i.e., along-shore velocity divided by U_0), U_0 is the dimensional velocity scale ($U_0 = (dh'/d\eta)L^2 f/D$), f is the Coriolis parameter, D is the dimensional depth scale (Fig. 11.1),

$$y = \delta\phi_2, \tag{11.1.4a}$$

$$z = \tfrac{1}{2}x^2 - x + \delta\phi_1, \tag{11.1.4b}$$

$$g_1 = -rx + \tau_0 + \tau(t), \tag{11.1.5a}$$

$$g_2 = -ry, \tag{11.1.5b}$$

$$g_3 = -rz - \tfrac{1}{2}rx^2 + (x-1)[\tau_0 + \tau(t)], \tag{11.1.5c}$$

$$H(x, y, z) = \tfrac{1}{2}y^2 + zx + \tfrac{1}{2}(\omega_0^2 - z)x^2 - \tfrac{1}{2}x^3 + \tfrac{1}{8}x^4, \tag{11.1.5d}$$

$$\omega_0^2 = 1 + \delta^2, \tag{11.1.5e}$$

$$\delta = \delta'/[\sqrt{2}R_0 D], \tag{11.1.5f}$$

$\epsilon r = \delta_E/2R_0D$ is the friction coefficient related to the eddy viscosity of the ocean flow, δ_E is the depth of the ocean's bottom Ekman layer, R_0 is the Rossby number U_0/fL, and $\epsilon\tau_0$ and $\epsilon\tau(t)$ are the normalized steady wind stresses (i.e., stresses $\epsilon\tau_0'$ and $\epsilon\tau'(t)$ divided by the wind stress scale $\tau_* = \rho U_0^2 D/L$). Note that the excitation is a state-dependent function of time (Eq. 11.1.5c).

11.2 WIND VELOCITY FLUCTUATIONS AND WIND STRESSES

In this section we present a model for the horizontal wind velocity fluctuations. A widely accepted empirical model is the van der Hoven (1957) spectrum, developed on the basis of wind speed measurements. The spectrum has three main frequency intervals: one interval containing a spectral peak with a period of about 4 days, a second interval, known as the spectral gap, extending over periods of about 5 h to 3 min and having negligible energy, and a third interval with a peak at a period of about 1 min. The third interval is of interest only in micrometeorological and structural engineering applications. It contains fluctuation components with negligible spatial coherence for separations in excess of 100 m, say (Simiu and Scanlan, 1996, p. 64). Because the geometric scale in the problem at hand is much larger

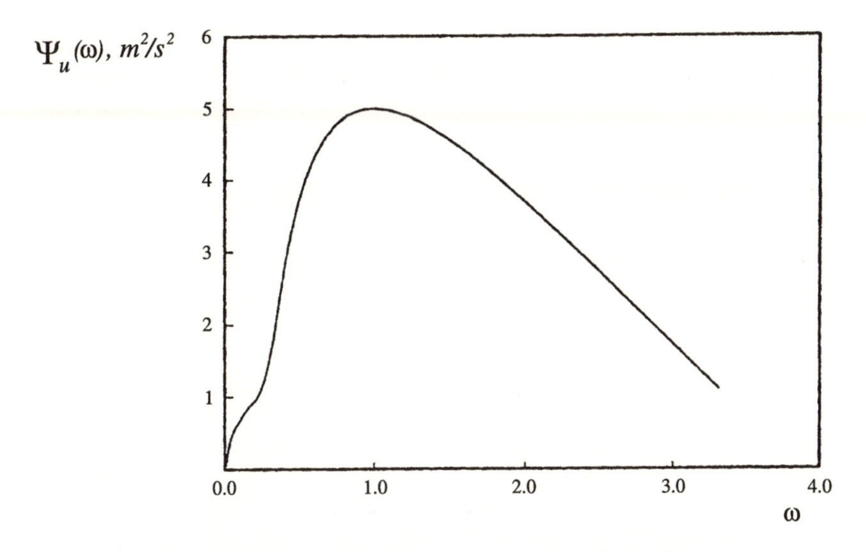

Figure 11.2. Spectral density of wind speed fluctuations.

than 100 m the effect of these components may be disregarded. The relevant portion of the spectrum is then, approximately,

$$\Psi_u(\omega) = \begin{cases} 0.2823\ln(\omega) + 1.300, & 0.01 \leq \omega \leq 0.10, \\ 0.4072\ln(\omega) + 1.599, & 0.10 \leq \omega \leq 0.30, \\ -2.71[\ln(\omega)]^2 + 5, & 0.30 \leq \omega \leq 3.85, \end{cases} \tag{11.2.1}$$

where $\omega = \Omega/\Omega_{\text{pk}}$, Ω is the dimensional frequency and $\Omega_{\text{pk}} \approx 2\pi/(4$ days) is the dimensional frequency corresponding to the spectral peak, which occurs at $\omega = 1$ (Fig. 11.2). The units of $\Psi_u(\omega)$ are m^2/s^2. For the spectrum of Eq. 11.2.1 the standard deviation of the wind velocity fluctuations $u(t)$ is $\sigma_u \approx 1.33$ m/s.

We assume that the effect of thermal stratification and of deviations from the prevailing wind direction is small. Also, on the basis of climatological data (e.g., NOAA, 1977), we assume that typical long-term mean wind speeds corresponding to the van der Hoven spectrum are of the order of $U = 6$ m/s. An approximate estimate of the coefficient of variation of the wind speed fluctuations is then $\sigma_u/U = 1.33/6.0 \approx 0.2$.

The total normalized wind stress $\epsilon\tau_0 + \epsilon\tau(t)$ is proportional to the square of the total wind speed $[U + u(t)]^2$; since $(\sigma_u/U)^2$ is small, on average the term $u^2(t)$ is small in relation to the fluctuating term $2Uu$ (Vaicaitis and Simiu, 1977; Simiu and Scanlan, 1996, p. 182). The spectral density of the fluctuations is therefore approximately proportional to $\Psi_u(\omega)$. We write $\epsilon\tau(t) \approx \epsilon\gamma G(t)$, where $G(t)$ has spectral density $\Psi_0(\omega) = \Psi_u(\omega)/\sigma_u^2$ and unit standard deviation, and $\epsilon\gamma$ is the standard deviation of the normalized fluctuating wind stresses. If we assume that the wind velocity fluctuations are Gaussian, it follows that surface stresses are approximately Gaussian.

11.3 DYNAMICS OF UNPERTURBED SYSTEM

For the unperturbed system ($\epsilon = 0$) the fixed points are yielded by the equations

$$y = 0, \tag{11.3.1}$$

$$x^3 - 3x^2 + 2(\omega_0^2 - z)x + 2z = 0 \tag{11.3.2}$$

obtained by equating to zero the right-hand sides of the unperturbed counterparts of (11.1.3). It can be verified that, for $z > z_c = \frac{3}{2}\delta^{4/3} + \delta^2 - \frac{1}{2}$, the phase plane diagram $x, y(z = \text{const})$ has a saddle point and two centers, and contains an asymmetrical pair of homoclinic orbits (Fig. 11.3). (For $z < z_c$ the phase plane diagram has one center and no saddle point.) The saddle point corresponds to the intermediate root of Eq. 11.3.2. Its coordinates are denoted by $x_s, y_s = 0$, and z_s. The saddle point and the homoclinic orbits depend continuously on z and form, respectively, a one-dimensional manifold $\Gamma(z)$ of (x, y, z) points and two-dimensional manifolds of (x, y, z) points similar to those represented partially in Fig. 2.13.

Let $x = x_s + x_h$, $x = y_s + y_h = y_h$, where x_h, y_h, are the coordinates with respect to the saddle point of the homoclinic orbit at elevation z_s. From the unperturbed equations of motion the following expression is obtained:

$$x_{h\pm}(t - t_0) = \frac{8k_s^2 x_{m\pm}}{bx_{m\pm}(1 + 1/\alpha_\pm)\cosh|k_s(t - t_0)|}, \tag{11.3.3}$$

$$k_s^2 = z_s - \omega_0^2 + 3x_s - (3/2)x_s^2 > 0, \tag{11.3.4}$$

$$\alpha_\pm = \frac{bx_\pm}{8k_s^2 - bx_{m\pm}}, \tag{11.3.5}$$

$$b = 4(x_s - 1) > 0, \tag{11.3.6}$$

$$x_{m\pm} = \tfrac{1}{2}[-b \pm (b^2 + 16k_s^2)^{1/2}] \tag{11.3.7}$$

(Allen et al., 1991). The coordinate $y_{h\pm}(t - t_0)$ of the homoclinic orbit at elevation z_s is obtained from the first of Eqs. 11.1.3, in which $\epsilon = 0$, and Eq. 11.1.5d:

$$y_{h\pm}(t - t_0) = \dot{x}_{h\pm}(t - t_0). \tag{11.3.8}$$

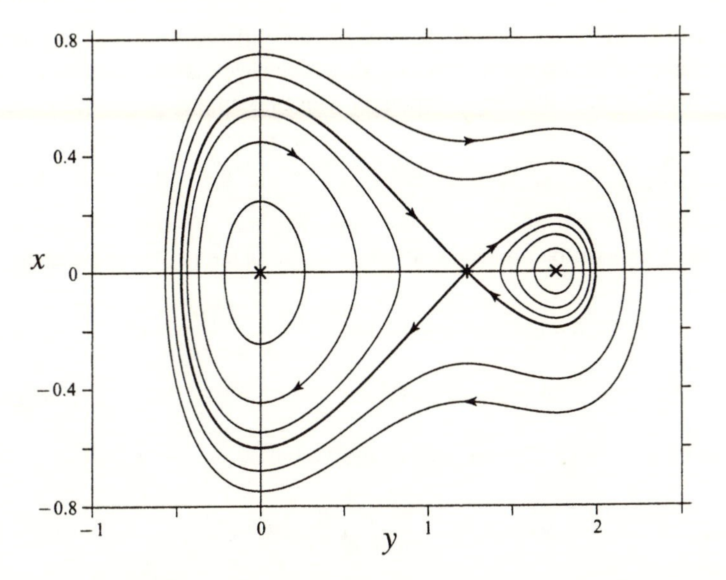

Figure 11.3. Homoclinic orbits and typical periodic orbits in unperturbed system (from Allen, Samelson, and Newberger, 1991).

11.4 DYNAMICS OF PERTURBED SYSTEM

It was shown in Chapter 2 that to find the hyperbolic orbit $\Gamma_\epsilon(z_0, t; \epsilon) = \Gamma(z_0) + O(\epsilon)$ in the perturbed system we need to use Eq. 2.8.10a. Averaging Eq. 11.1.5c with respect to time eliminates the fluctuating stress term, so Eq. 2.8.10a yields

$$z_0 = -\tfrac{1}{2}x_0^2 + (x_0 - 1)\tau_0/r. \tag{11.4.1}$$

The coordinates x_0, y_0, z_0 defining to first order the hyperbolic orbit $\Gamma_\epsilon(z_0, t; \epsilon)$ are obtained by substituting Eq. 11.4.1 in Eq. 11.3.2. The result is

$$x_0^3 - (2 + \tau_0/r)x_0^2 + (\omega_0^2 + 2\tau_0/r)x_0 - \tau_0/r = 0. \tag{11.4.2}$$

By Eq. 11.3.1 $y_0 = 0$, and z_0 is obtained by substituting the solution of (11.4.2) in Eq. 11.4.1. It is shown by Allen et al. (1991) that

$$d\overline{[g_3(\gamma(z))]}/dz\big|_{z_0} = -r[2k_0^2 + (x_0 - 1)^2 + \delta^2]/k_0^2 < 0, \tag{11.4.3}$$

where k_0 is the value of k_s at $z_s = z_0$. Therefore an orbit with initial conditions on $\Gamma_\epsilon(z, t; \epsilon)$ converges on $\Gamma_\epsilon(z_0, t; \epsilon)$, as in Fig. 2.14.

Calculations based on Eq. 2.8.14 yield the following expressions for the expectation of the Melnikov process rC_1 and the Melnikov scale factor $C_2(\omega)$ corresponding to a harmonic excitation with unit amplitude:

$$rC_1 = rC_1^{\pm} = r(x_0 - \tau_0/r)[8d \tan^{-1}(x_{m\pm}/(2k_0)) - k_0 b], \tag{11.4.4}$$

where the plus and minus signs correspond to the left-hand side and right-hand side unperturbed homoclinic orbits in the plane (x, y), respectively,

$$C_2(\omega) = C_2^{\pm}(\omega) = -4\pi d \frac{\sinh[\omega \cos^{-1}(\alpha_{\pm})/k_0]}{\sinh(\omega\pi/k_0)} \tag{11.4.5}$$

$$d = k_0^2 + (x_0 - 1)^2 \tag{11.4.6}$$

(Allen et al., 1991). The variance of the Melnikov process is (Eq. 4.1.17b)

$$\mathrm{Var}[M(t)] = \frac{\gamma^2}{2\pi} \int_0^{\infty} C_2^2(\omega) \Psi_0(\omega)\, d\omega. \tag{11.4.7}$$

11.5 NUMERICAL EXAMPLE

We consider the case, studied for harmonic excitation by Allen et al. (1991), $\delta = 0.3003, \tau_0/r = 3.236$. It can be verified that the fixed points of the averaged system are $\{0,0,0\}, \{1.236,0,0\}$, and $\{1.764,0,0\}$ (Eqs. 11.3.1, 11.4.2, and 11.4.1). From Eqs. 11.4.4 and 11.4.5, $C_1^+ = 2.524, C_1^- = -7.076$ (the $-$ and $+$ superscripts correspond to the left- and right-hand wells) and

$$C_2^+(\omega) = -4.8 \sinh(2.064\omega)/\sinh(5.5\omega), \tag{11.5.1}$$

$$C_2^-(\omega) = -4.8 \sinh(3.436\omega)/\sinh(5.5\omega). \tag{11.5.2}$$

For *harmonic forcing* with $\omega = 1$ (a case examined by Allen et al. (1991) and assumed therein to correspond to a dimensional time $T_{\mathrm{pk}} \approx 4$ days), we have $C_2^+ = -0.152, C_2^- = -0.609$. Assuming that the forcing is $\epsilon\tau(t) = \epsilon\sqrt{2}\gamma \cos(\omega t)$ (i.e., its standard deviation is $\epsilon\sqrt{2}\gamma$), the necessary condition for escapes from the left well is satisfied for $\gamma/r > 8.22$.

We now consider the case of *random forcing* with spectrum $\epsilon^2\gamma^2\Psi_0(\omega)$, where $\epsilon\gamma$ now denotes the standard deviation of the stochastic forcing. Figures 11.4a and 11.4b show, respectively, the square of the Melnikov scale factor $[C_2^-(\omega)]^2$ and the spectral density of the Melnikov process $\Psi_{M_{0-}}(\omega) = [C_2^-(\omega)]^2 \Psi_0(\omega)$; Figs. 11.5a and 11.5b represent $[C_2^+(\omega)]^2$ and $\Psi_{M_{0+}}(\omega) = [C_2^+(\omega)]^2 \Psi_0(\omega)$, respectively. The factors $C_2(\omega)$ are seen to suppress or considerably reduce the spectral components of the wind stress with frequencies $\omega > 1.5$ or so, and to amplify lower-frequency components.

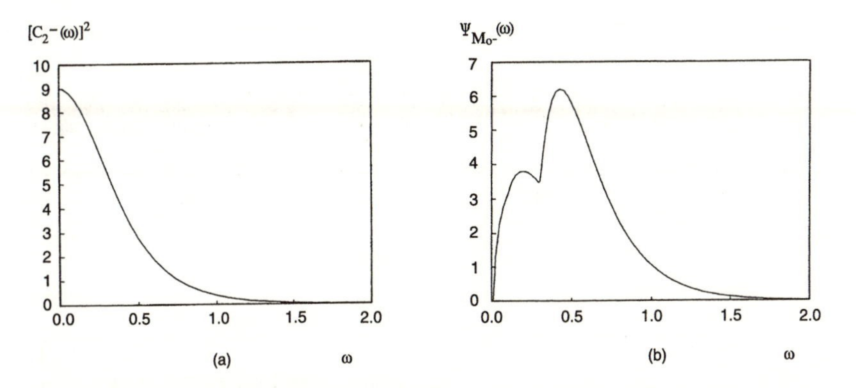

Figure 11.4. (a) Square of Melnikov scale factor $[C_2^-(\omega)]^2$, and (b) spectral density $\Psi_{M_{0-}}(\omega)$ (from Simiu, 1996).

Equation 5.5.3a applied to $\Psi_{M_{0-}}(\omega)$ and $\Psi_{M_{0+}}(\omega)$ yields $\nu^-=0.0702$ and $\nu^+=0.0534$, respectively. The standard deviations of the Melnikov process are $\sigma_{M^+}=0.359\gamma$ and $\sigma_{M^-}=0.778\gamma$, respectively, so $\kappa_-=|E[M_-]/\sigma_{M^-}|=9.1r/\gamma$ and $\kappa_+=|E[M_+]/\sigma_{M^+}|=7.03r/\gamma$ (Eq. 5.5.3b).

Recalling that in the harmonic excitation case no chaos is possible for $\gamma/r<8.22$, we consider the case $\gamma/r=4$, so $\kappa_-=2.273$ and $\kappa_+=1.76$. From Eq. 5.5.2, $\tau_u(\kappa_-)=188.6$ (since $\omega=1$ corresponds to a period of 4 days, the dimensional time between consecutive zero upcrossings of the Melnikov process is $188.6\times4/(2\pi)=119.7$ days) and $\tau_u(\kappa_+)=88.5$ (56.3 days), the left well being larger than the right well (see Fig. 11.3). A lower bound for the probability that no escapes will occur from the left well during a time interval $T=1$ month (i.e.,$2\pi(30\text{days})/(4\text{days})=47.1$ nondimensional time

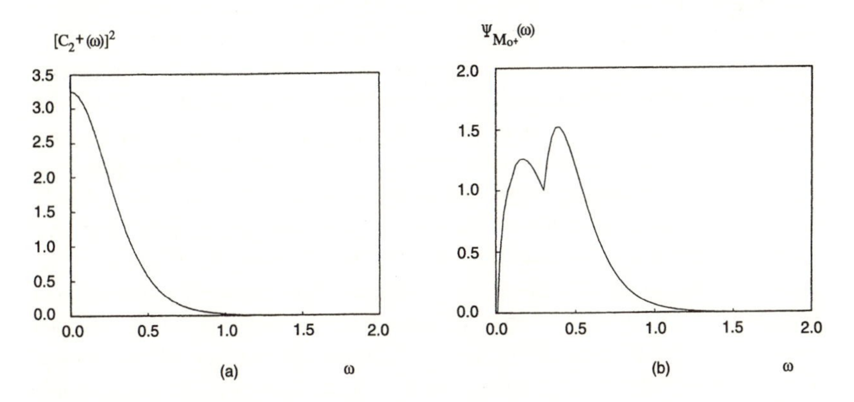

Figure 11.5. (a) Square of Melnikov scale factor $[C_2^+(\omega)]^2$, and (b) spectral density $\Psi_{M_{0+}}(\omega)$ (from Simiu, 1996).

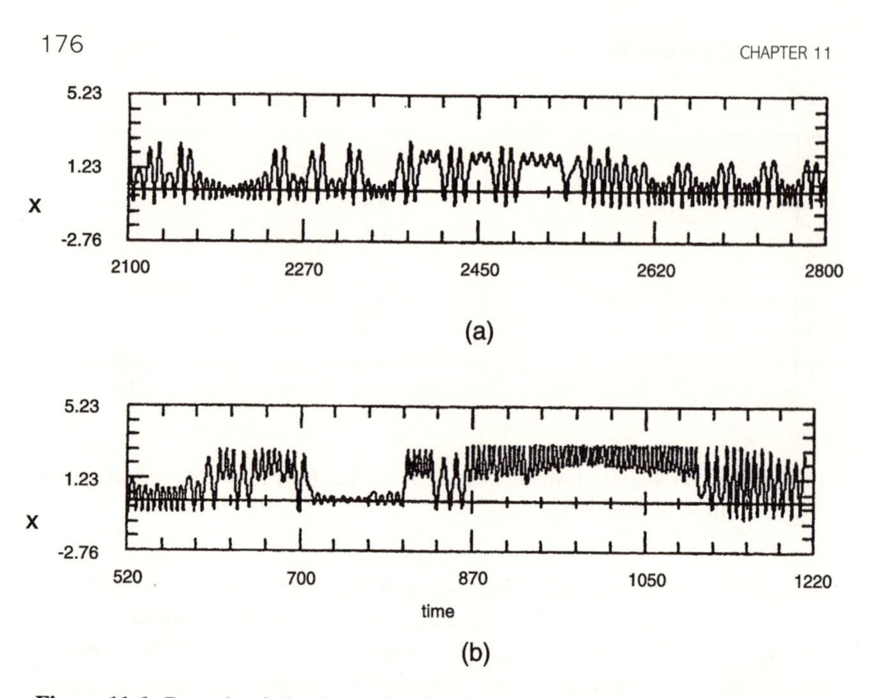

Figure 11.6. Records of chaotic motion for (a) harmonic forcing; (b) realization of stochastic forcing.

units) is $p_M(0, T^- = 1 \text{ month}) = 0.78$ (Eq. 5.5.4). For the right well p_M (0, $T^+ = 1$ month) $= 0.59$.

To recapitulate, the possibility of the occurrence of escapes associated with chaotic motions was assessed using the Melnikov approach for two cases with identical parameters, but different types of fluctuation of the wind-induced excitation. In the first case the fluctuations were harmonic with standard deviation $\gamma = 8.22r$ and frequency $\omega = 1$ (dimensional period $T = 4$ days), that is, the entire energy of the fluctuations was concentrated at the frequency of the wind stress spectral peak. From the Melnikov approach it follows that no escapes are possible in this case. In the second case the fluctuations were random and were derived from the van der Hoven spectrum whose peak is at $T_{\text{pk}} = 4$ days. The standard deviation of the wind stresses was $\gamma = 4r$, that is, less than half as large as that of the largest harmonic forcing consistent with the nonoccurrence of escapes ($\gamma = 8.22r$). In this case the necessary condition for the occurrence of chaos is satisfied. The probability that escapes associated with chaotic motions can occur from the larger potential well was estimated to be at most $1 - 0.78 = 0.22$ during an interval of one month. The upper bounds of the probabilities that such escapes are possible would increase considerably if the standard deviation of the wind stresses were assumed to be $\gamma = 8.22r$, rather than just $\gamma = 4r$. We recall, however, that a complete study of escapes would also include possible escapes of a non-chaotic nature induced by either deterministic or stochastic excitation—see Section 2.8.3.

Illustrations of chaotic time histories of $x(t)$ for $\delta=0.3003, \epsilon=0.001, \tau_0/r=3.236$, and $r=0.01$ are shown in Figs. 11.6a and 11.6b for, respectively, periodic forcing, $\gamma=21.21, \omega=1, x(0)=1.236$, $y(0)=z(0)=0$, and stochastic forcing, $\gamma=8$, $x(0)=10^{-5}$, $y(0)=z(0)=0$. Sensitivity to initial conditions was verified numerically by following the evolutions in time of small separations introduced in the initial values of the equations of motion, using the procedure illustrated in Fig. 3.3.

Chapter Twelve

The Auditory Nerve Fiber as a Chaotic Dynamical System

The auditory nerve fiber is a natural device of interest to both neurophysiologists and signal processing engineers. Its dynamics consists of random low-amplitude motions from which escapes occur at irregular intervals. The escapes are referred to as *firings*, and are associated with random, high-amplitude bursts called *spikes*. A simulated time history of such motions is shown in Fig. 12.1.

In this chapter we review results of three sets of experiments on the behavior of the auditory nerve fiber. The results strongly suggest the modeling of the fiber as a chaotic, one-degree-of-freedom dissipative system with an asymmetrical double-well potential, and are explained by Melnikov theory in a remarkably consistent manner. We also reproduce results of simulations based on the classical Fitzhugh-Nagumo (F-N) model of the auditory nerve fiber. Unlike the proposed chaotic model, the F-N model appears to be inconsistent with the experimental results.

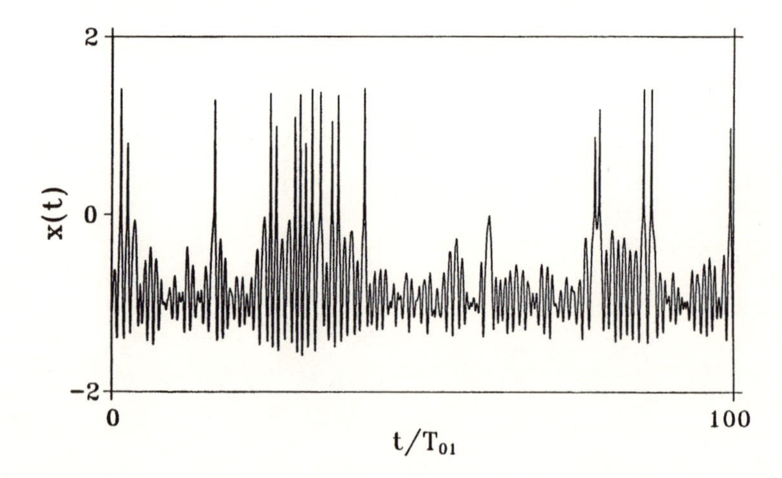

Figure 12.1. Simulated time history of motion induced by harmonic excitation with period T_{01} in the presence of weak noise.

Section 12.1 reviews the neurophysiological experimental results and discusses their consistency with a model to which Melnikov theory applies. Section 12.2 shows results of simulations based on the F-N model, and their disagreement with the experimental observations. Section 12.3 describes the proposed model in detail. Section 12.4 shows results of numerical simulations based on the model, and compares them with experimental results. Section 12.5 contains concluding remarks.

12.1 EXPERIMENTAL NEUROPHYSIOLOGICAL RESULTS

This section presents experimental results pertaining to (a) nerve fibers subjected to harmonic excitation in the presence of spontaneous activity (i.e., of weak, random motion, observed even in the absence of experimentally induced excitation, and that may be assumed for modeling purposes to be caused by weak noise); (b) nerve fibers excited by broadband white noise excitation; (c) nerve fibers excited by a sum of two harmonic terms in the presence of spontaneous activity. Features of the experimental results that motivate the use of a Melnikov-based model are pointed out for each of the three sets of experiments.

12.1.1 Fibers Subjected to Harmonic Excitation in the Presence of Spontaneous Activity

Experiments reported by Rose et al. (1967) show that (1) for fixed excitation amplitude, mean firing rates are highest for frequencies contained in a relatively narrow "best" interval, and decrease for frequencies outside that interval until, for both low and high frequencies, they become vanishingly small (the term "best" interval is commonly used in the neurophysiological literature). In addition, (2) for fixed excitation frequency, the mean firing rates increase as the excitation amplitude increases (Fig. 12.2).

These two qualitative features are guaranteed to occur in a wide class of chaotic systems with double-well potential. Recall that, for such systems, a harmonic excitation induces a Melnikov function with nonzero mean and amplitude proportional to the Melnikov scale factor (see, e.g., Eq. 2.5.7). The "best" frequencies correspond to the frequencies centered around the frequency of the Melnikov scale factor's peak (Section 5.6). These frequencies are the most effective, in the sense that they correspond to excitations for which the system's mean escape time is lowest (i.e., the system's mean escape rate is highest).

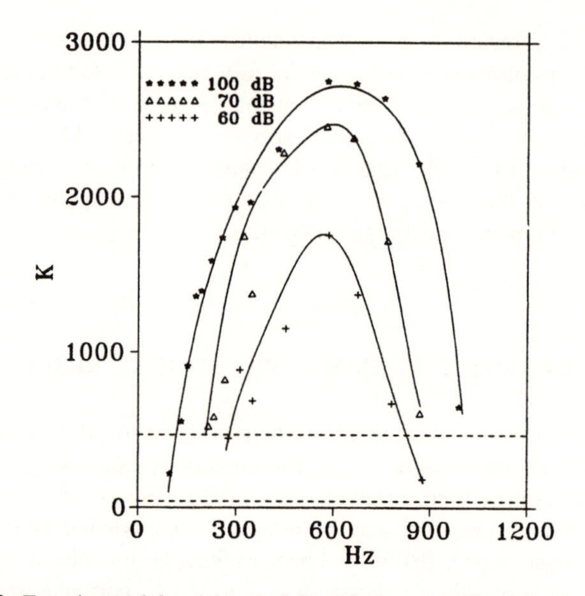

Figure 12.2. Experimental dependence of K on frequency and amplitude of harmonic excitation. K is the number of spikes observed in 20 s for harmonically excited auditory nerve fiber in squirrel monkeys. Dashed lines indicate levels of spontaneous activity (after Rose et al., 1967).

12.1.2 Nerve Fibers Subjected to White Noise Excitation

Experiments on auditory nerve fibers excited by nearly white noise show that interspike interval histograms[1] (ISIH's) are multimodal. The time interval between successive histogram peaks (modes) is approximately equal to the "best" period, that is, the period corresponding to the nerve fiber's "best" frequency—see Fig. 12.3 (Ruggero, 1973).

Like the experimental results for harmonically excited fibers, this result can be explained qualitatively in terms of the proposed bistable model's behavior. For a system excited by white noise, the Melnikov scale factor filters out components with ineffective frequencies, and the system responds predominantly to excitations with frequencies at or near the "best" frequency. Therefore, the spectrum of the response is qualitatively similar to the spectrum of the response due to harmonic excitation: it has, in addition to broadband frequency components, a peak at or near the "best" frequency, and subsidiary peaks at or near the frequencies of the corresponding subharmonics. For

[1] ISIH's plot numbers of spikes as functions of the time interval separating them. For example, for a periodic function with period T the ISIH is unimodal and consists of the Dirac delta function $\delta(t - T)$.

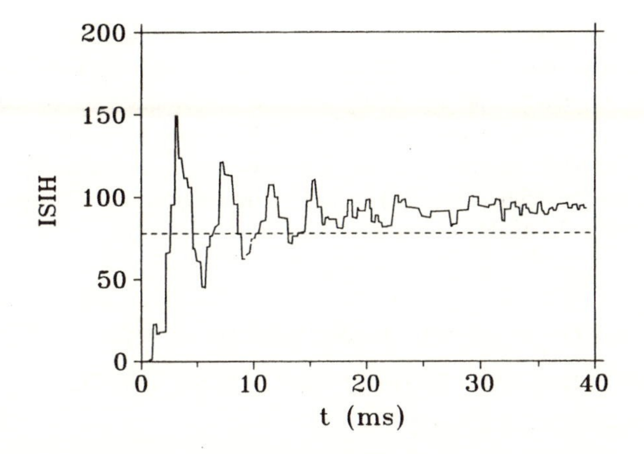

Figure 12.3. Interspike interval histogram for response to white noise in experiments on auditory nerve fiber in squirrel monkey. Dashed line indicates level of spontaneous firing rate. Modes are approximately equal to integer multiples of the "best" period (after Ruggero, 1973).

the system excited by white noise the effective excitation components are spread over a relatively narrow band of frequencies, while for the harmonically excited system the spectral peaks induced by the excitation are sharp. This difference is inconsequential qualitatively, although quantitatively it is reflected in the spread exhibited by the ISIH's around the frequencies of the modes—see Fig. 12.3.

12.1.3 Nerve Fibers Excited by a Sum of Two Harmonic Terms in the Presence of Spontaneous Activity

Results of experiments in which auditory nerve fibers were excited by two harmonic functions in the presence of spontaneous activity were reported by Hind et al. (1967). In those experiments the frequency of one of the harmonic excitations, denoted by ω_{01}, was close to the "best" frequency ω_{best}, while for the second harmonic $\omega_{02} < \omega_{\text{best}}$, so that its effectiveness in inducing firing was relatively weak. The experimental results showed that the ISIH's are multimodal, with basic period $T_{01} = 2\pi/\omega_{01}$ or $T_{02} = 2\pi/\omega_{02}$ according as the ratio $\eta = \gamma_{01}/\gamma_{02}$ of the amplitudes corresponding to ω_{01} and ω_{02} is relatively large or small.

These results are qualitatively consistent with the behavior of the proposed bistable model for the following reasons. If η is large the excitation harmonic with effective frequency ω_{01} (i.e., with frequency at or close to the Melnikov scale factor's peak) is dominant. The effect of the second harmonic is marginal, both because its frequency is ineffective and because its amplitude is small. The behavior of the proposed model is therefore qualitatively similar to the behavior noted earlier for the case of harmonic excitation, in which the spectrum of the system response has a relatively large peak at the excitation frequency, and subsidiary peaks at the frequencies of the subharmonics. This translates into multimodal ISIH's with basic period T_{01}. If η is small, the large amplitude of the harmonic term with frequency ω_{02} can make up for the relative ineffectiveness of that frequency. The basic period of the multimodal ISIH is then T_{02}.

12.2 RESULTS OF SIMULATIONS BASED ON THE FITZHUGH-NAGUMO MODEL. COMPARISON WITH EXPERIMENTAL RESULTS

According to Hochmair-Desoyer et al. (1984), in the absence of noise the Fitzhugh-Nagumo model predicts correctly that, as the excitation frequency increases beyond the "best" frequencies, the amplitude of the harmonic signal needed to cause firing increases sharply. However, it appears that the model fails to predict a similarly sharp increase for excitation frequencies lower than the "best" frequencies (Fig. 12.4). In the presence of noise the disagreement between typical F-N model predictions and the experimental results of Rose et al. (1967) is even stronger (Hochmair-Desoyer et al., 1984, p. 561).

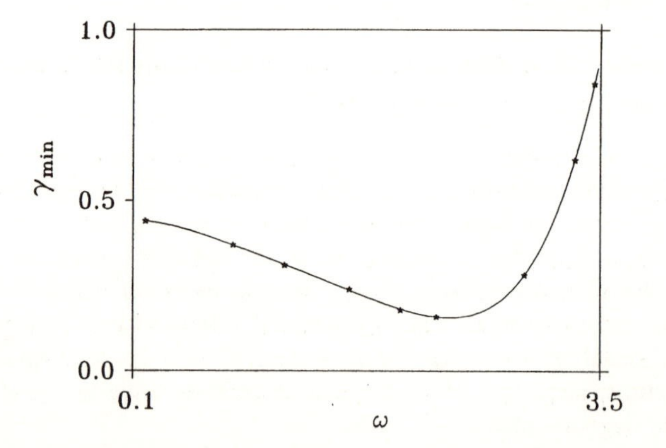

Figure 12.4. Dependence of minimum amplitude needed to produce firings, γ_{min}, on frequency of the harmonic excitation in the absence of noise, as modeled by Fitzhugh-Nagumo equation. Horizontal scale is logarithmic (after Hochmair-Desoyer et al., 1984). γ_{min} increases sharply for high frequencies, in accordance with the experimental results of Fig. 12.1. However, the model fails to reflect a similar increase for low frequencies.

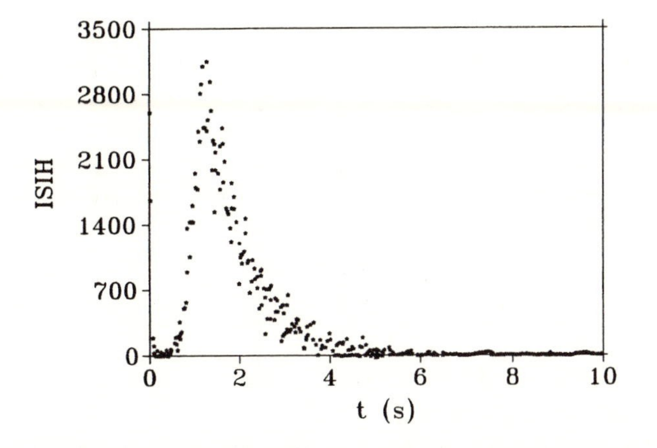

Figure 12.5. Interspike interval histogram for response induced by white noise as simulated by the Fitzhugh-Nagumo equation. The histogram is unimodal, unlike the experimentally obtained histogram of Fig. 12.2 (after Longtin, 1993).

According to Longtin (1993), for white noise excitation the F-N model yields a unimodal ISIH (Fig. 12.5), in disagreement with the experimental results of Ruggero (1973) (Fig. 12.3).

12.3 ASYMMETRIC BISTABLE MODEL OF AUDITORY NERVE FIBER RESPONSE

The proposed bistable model for the response of the auditory nerve fiber is

$$\ddot{x} = -V'(x) + \epsilon[\gamma_{01}\cos(\omega_{01}t) + \gamma_{02}\cos(\omega_{02}t) + \sigma G(t) - \beta\dot{x}], \quad (12.3.1)$$

where $V(x)$ is a double-well potential defined by a polynomial, ϵ is a perturbation parameter that may have a stepwise variation over x, $G(t)$ is a nearly white noise process, and σ, β are adjustable parameters. Additional adjustable parameters of the model are those defining the polynomial $V(x)$, and the constants defining the stepwise variation of ϵ. Equation 12.3.1 with the appropriate choice of parameters (1) can achieve simulations of typical nerve fiber time histories (i.e., time histories with irregular low-amplitude motions interrupted by irregular, high-amplitude spikes), such as the plot of Fig. 12.1; (2) has solutions with the qualitative features observed in the experiments described earlier. In addition, Eq. 12.3.1 is capable of reproducing quantitatively the behavior observed in any given experiment.

We first show that the system 12.3.1 must be asymmetric with respect to the $\dot{x}$ axis. The asymmetry involves (1) the double-well potential, which is deeper for $x > 0$ than for $x \leq 0$, as shown in Fig. 12.6a; and (2) the

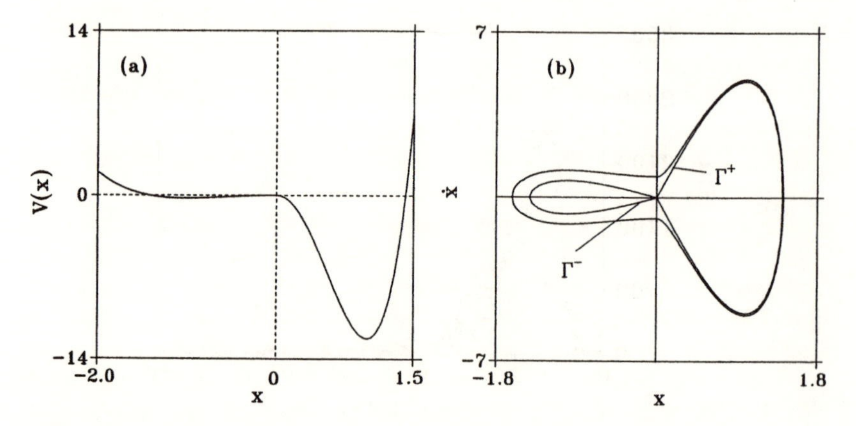

Figure 12.6. (a) Potential $V(x)$; (b) phase plane diagram showing homoclinic orbits and an orbit visiting the half-planes $x \leq 0$ and $x > 0$ (after Franaszek and Simiu, 1998).

system's perturbation, which in the proposed model is defined by parameter values $\epsilon > 0$ for $x \leq 0$ and $\epsilon = 0$ for $x > 0$ (i.e., the perturbation vanishes for $x > 0$). Note that, with no loss of generality, we may choose $\epsilon = 1$ for $x \leq 0$. This merely affects the choice of the parameters γ_{01}, γ_{02}, σ, and β.

Associated with the unperturbed system ($\epsilon = 0$ for all x) are the homoclinic orbits Γ^- and Γ^+ shown in Fig. 12.6b. We require that the low-amplitude response of the system consist of chaotic motions within the inner core associated with the homoclinic tangle that represents the perturbed counterpart of the homoclinic orbit Γ^-. For any specified excitation $\epsilon[\gamma_{01}\cos(\omega_{01}t) + \gamma_{02}\cos(\omega_{02}t) + \sigma G(t)]$, $\epsilon\beta$ must therefore be chosen so that the left-hand side of the Melnikov inequality representing the necessary condition for chaos in the half-plane $x \leq 0$,

$$-4\beta/3 + \gamma_{01}S(\omega_{01}) + \gamma_{02}S(\omega_{02}) + \sigma\sum_{k=1}^{K} a_k S(\omega_k) > 0, \qquad (12.3.2)$$

be sufficiently large so that chaos occurs. (In Eq. 12.3.2 a_k are the amplitudes of the harmonic components in the approximation $G_N(t)$ of the process $G(t)$, see Eq. 4.2.5, and $S(\omega)$ is the Melnikov scale factor corresponding to a unit amplitude of a harmonic excitation.) The fact that the motion is chaotic allows it to escape from the inner core through the mechanism of chaotic transport across the pseudoseparatrix, that is, the motion can follow a trajectory similar to the curve shown next to the homoclinic orbit Γ^- in Fig. 12.6b. After reaching the coordinate $x = 0$ the motion continues in the half-plane $x > 0$. Since in this half-plane the perturbation vanishes ($\epsilon = 0$), the motion cannot be chaotic, and no crossing can therefore occur into the inner core defined by the homoclinic orbit Γ^+. Rather, a large-amplitude motion close to Γ^+ occurs that returns the trajectory to the half-plane $x \leq 0$, where it again becomes chaotic and either stays outside the pseudoseparatrix or is entrained into the inner core (see Section 2.7.2). Because the motion for $x \leq 0$ is chaotic, each trajectory that intersects the axis $\dot{x}$ does so at a different point, so that no two trajectories near Γ^+ are the same.

Having discussed the qualitative considerations that dictate the asymmetry in the choice of the parameter ϵ, we now turn to the asymmetry of the double-well potential. For fixed maximum homoclinic orbit coordinate $x_{\max} \neq 0$ such that $V(x_{\max}) = 0$, the deeper a potential well, the larger are the velocities in trajectories close to its homoclinic orbit (the velocity being equal to the ordinate of the trajectory in the phase plane—see Fig. 12.6b). The choice of the depth of the well for the half-plane $x > 0$ is dictated by the need to achieve a relatively small time of travel for the motions in that half-plane. Had the well for $x > 0$ been much shallower than in Fig. 12.6a, the time intervals between two successive minima of a spike would have been much longer than in Fig. 12.1.

Although potential functions with various polynomial forms may be used, it is reasonable, at least to a first approximation, to try the potential

$$V(x) = \alpha^-(-x^2/2 + x^4/4), \qquad x \leq 0, \tag{12.3.3a}$$

$$V(x) = \alpha^+(-x^2/2 + x^4/4), \qquad x > 0, \tag{12.3.3b}$$

which represents an asymmetric, modified version of the Duffing–Holmes potential (Eq. 2.5.14c). The coordinate $x_{\max}$ is independent of $\alpha^\pm$, on the other hand, the larger $\alpha^\pm$, the deeper is the well and the larger are the velocities on and near the homoclinic orbit (see Eq. 2.5.16, where x_{h2} represents the velocity of a point on the homoclinic orbit). It can be verified immediately that $V'(0) = 0$. For $x \leq 0$, the Melnikov scale factor corresponding to a harmonic excitation with unit amplitude (see Remark at the end of Section 2.5.3 and following Eq. 5.2.6) is

$$S(\omega) = (2/\alpha^-)^{1/2}\pi\omega \operatorname{sech}\{\pi\omega/2\sqrt{\alpha^-}\} \tag{12.3.4}$$

(Eq. 2.5.17). Equation 12.3.4 is represented in Fig. 12.7 for $\alpha^- = 1$.

We note that the dimensional counterparts of the terms $\ddot{x}$ and $\epsilon\gamma_{01} \times \cos(2\pi f_{01}t)$ ($\omega_{01} = 2\pi f_{01}$) in Eq. 12.3.1 are, respectively, $d^2X/d\tau^2$ and $\epsilon A P_{01}\cos(2\pi F_{01}t)$, where $X = c_1 x$, $\tau = c_2 t$, $F_{01} = f_{01}/c_2$, X, τ, and A have dimensions mV, ms, and mV ms^{-2}, respectively, and P_{01} is expressed in dB. Thus, $\gamma_{01} = A c_2^2 P_{01}/c_1$. Typical amplitudes of firings and "best" frequencies reported by Rose et al. (1967) for the experiments of Fig. 12.2 were about 1 mV and 600 Hz, respectively.

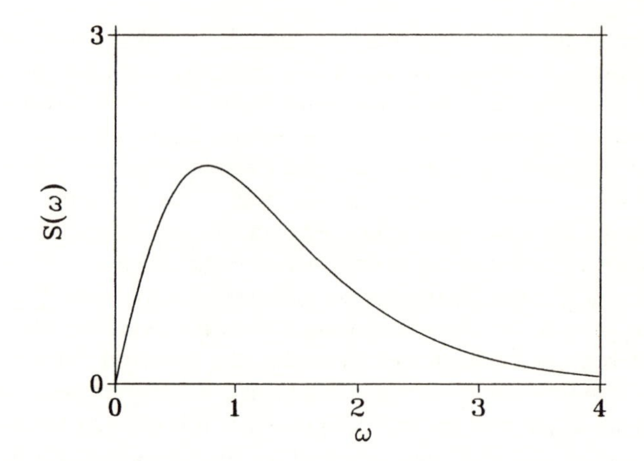

Figure 12.7. Melnikov scale factor (Eq. 12.3.3, or Eq. 2.4.23 in which $\gamma = 1$), for $\alpha^- = 1$.

12.4 NUMERICAL SIMULATIONS

The simulations aimed at reproducing qualitatively observed fiber behavior (Fig. 12.2) were based on the assumption that the potential function is described by Eqs. 12.3.3. The simulation of white noise was based on the Bennett-Rice representation (Eq. 4.2.5) and the one-sided spectral density

$$\Psi_0(\omega) = 2/(1 + c^2\omega^2) \tag{12.4.1}$$

in which $c = 0.02$, that is, the spectral density varies slowly with frequency and is therefore a close approximation of white noise (see Section 4.2.3.2).

All the results for which simulations were performed were structurally stable (robust), that is, no changes in qualitative behavior occurred when the system parameters were varied within reasonably wide intervals. We remark that robustness is implicit in the fact that, for chaotic behavior to occur, the realization of the Melnikov process must have simple zeros. If it does have simple zeros, for a sufficiently small change in the system parameters it will still have simple zeros: such a change will not result in a different qualitative behavior of the system; in other words, the system is robust.

12.4.1 Harmonic Excitation in the Presence of Weak White Noise

Consider an excitation defined by the parameter values $\gamma_{01} = 0.12$, $\gamma_{02} = 0$, $\omega_{01}/\omega_{\text{best}} = 1.0$, $\sigma = 0.005$, and the spectral density of the process $G(t)$ given by Eq. 12.4.1 in which $c = 0.02$. The time history shown in Fig. 12.1 is based on the parameters, obtained by trial and error, $\alpha^- = 1$, $\alpha^+ = 49$

(to which there correspond the wells shown in Fig. 12.6a), and $\beta = 0.16$. With this excitation, potential function, and damping parameter the motion of the system is chaotic. Note the presence in Fig. 12.1 of both single and multiple firings (a double firing can be seen at time $t/T_0 \approx 80$). This is consistent with experimental results noted by Rose et al. (1967, p. 776).

12.4.1.1 Dependence of the Mean Firing Rate on Frequency and Amplitude of Harmonic Excitation

Based on the results of Section 5.6 we expect that for finite, as opposed to asymptotically small, perturbations the dependence of the model response on the Melnikov scale factor shown in Fig. 12.7 is maintained in a qualitative sense, and that response patterns are therefore similar to those obtained experimentally (Fig. 12.2). Simulations were performed with the following parameters: $\gamma_{01} = 0.1$ and $\gamma_{01} = 0.12$, $\gamma_{02} = 0$, $\sigma = 0.005$, $\alpha^- = 1$, $\alpha^+ = 49$, $\beta = 0.16$, and $c = 0.02$ (Eq. 12.4.1).

Figure 12.8 shows the dependence of the mean firing rate r on frequency $\omega_{01}/\omega_{\text{best}}$ for the two amplitudes of the harmonic forcing. For both amplitudes the harmonic excitation is most effective in producing firings for frequencies close to the "best" frequency, and is increasingly ineffective as the excitation

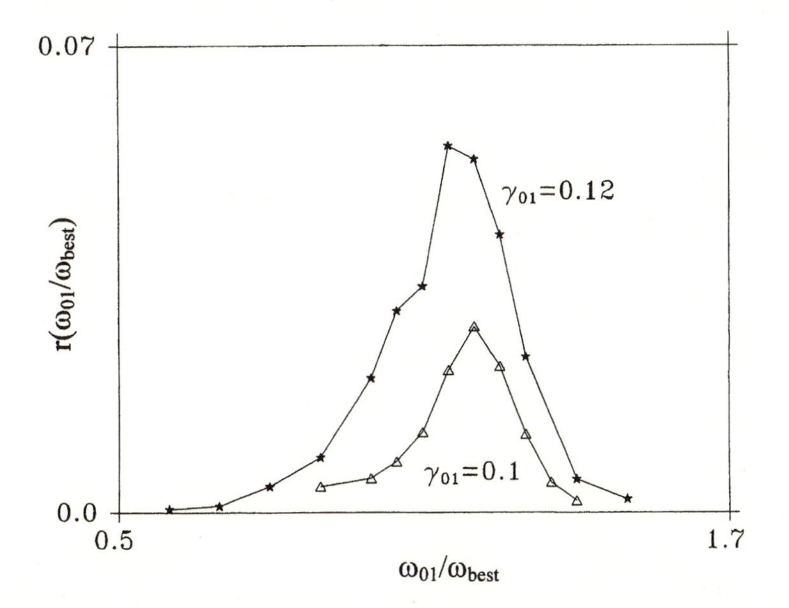

Figure 12.8. Dependence of firing rate r on harmonic excitation frequency for two excitation amplitudes γ_{01}, simulated by Eq. 12.3.1 with harmonic excitation in the presence of noise. The dependence of r on excitation frequency and amplitude is qualitatively similar to that observed in the experiments of Fig. 12.2.

frequencies—either low or high—are farther away from the "best" frequency. For fixed frequency, the larger amplitude causes a larger firing rate. These qualitative results are inherent in the structure of the bistable model. Quantitative results depend upon the choice by trial and error of the polynomial potential and the dissipation parameter β.

12.4.1.2 Interspike Interval Histograms for Harmonic Excitation in the Presence of Weak White Noise

We assume again $\alpha^- = 1$, $\alpha^+ = 49$, $\beta = 0.16$, $\gamma_{02} = 0$, $\sigma = 0.005$, and the spectral density of $G(t)$ given by Eq. 12.4.1, $c = 0.2$. Figures 12.9a–c show ISIH's for $\gamma_{01} = 0.12$ and $\omega_{01}/\omega_{\text{best}} = 1.25, 1.05$, and 0.8, respectively, that is, they correspond, respectively, to harmonic excitation frequencies higher than, approximately equal to, and lower than the "best" frequency. The period of the harmonic excitation is $T_{01} = 2\pi/\omega_{01}$. The ISIH's are multimodal and agree qualitatively with ISIH's based on experiments (Rose et al., 1967). Both for the experiments and for Figs. 12.9a–c the peaks in the ISIH's are grouped around the integer multiples of the period of the harmonic excitation. The

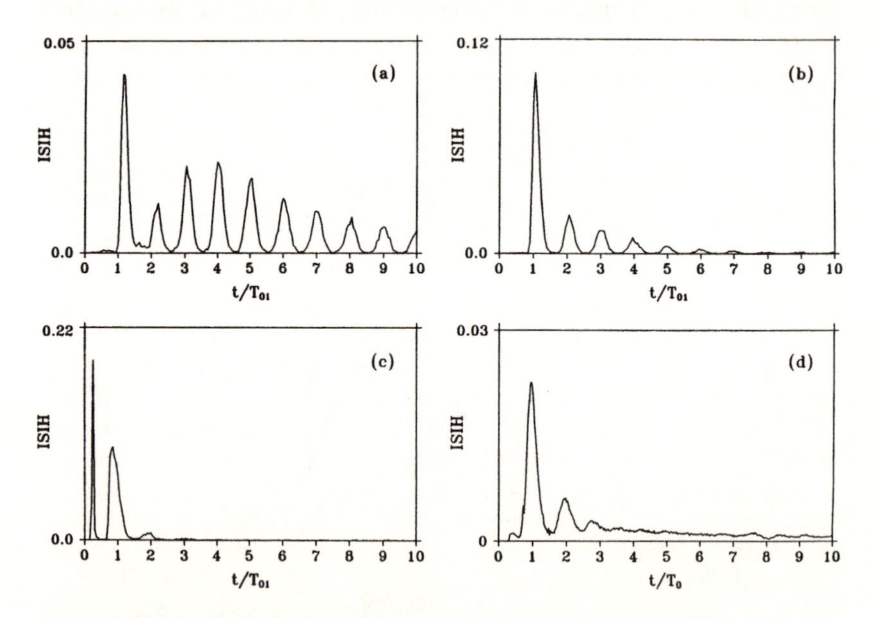

Figure 12.9. Interspike interval histograms simulated from Eq. 12.3.1: (a–c) Harmonic excitation in the presence of noise for excitation frequency larger than, approximately equal to, and smaller than the "best" frequency, respectively; (d) excitation by nearly white noise. T_{01} and T_0 denote the period of the harmonic excitation and the period corresponding to the "best" frequency, respectively.

firing is aperiodic, however, which is consistent with the fact that the motions are chaotic. As noted earlier, the preference for the forcing period and its integer multiples reflects the large spectral ordinate of the response at the forcing frequency and the frequencies of the corresponding subharmonics. Note the presence in Figs. 12.9a,b,c of components with periods shorter than the dominant period. As was pointed out by Rose et al. (1967, p. 776), such components reflect the existence of multiple firings.

12.4.2 Interspike Interval Histograms: Excitation by White Noise

It was noted earlier that, for white noise excitation, the F-N model does not agree with the experimental results of Ruggero (1973): while the ISIH's obtained from experiments are multimodal, those based on the F-N model were found to be unimodal, as shown in Fig. 12.5 (Longtin, 1993). In contrast, simulations based on Eq. 12.3.1 agree well with the experimental results of Ruggero (1973). This is illustrated by Fig. 12.9d, which shows a typical ISIH based on Eq. 12.3.1 with the same parameters as those of Figs. 12.9a–c, except that $\gamma_{01} = 0$ and $\sigma = 0.035$, that is, the only excitation is stochastic. The ISIH is multimodal with periodicities closely related to the "best" frequency. The results of Fig. 12.9d are consistent with Melnikov theory since components with frequencies equal and close to the "best" frequency—the frequency of the Melnikov scale factor's peak—are the most effective in inducing escapes. The behavior of the system can therefore be expected to be determined by those components and their subharmonics.

12.4.3 Interspike Interval Histograms: Excitation by a Sum of Two Harmonics in the Presence of Weak White Noise

The experiments by Hind et al. (1967) discussed earlier involved a harmonic excitation with frequency ω_{01} close to the "best" frequency ω_{best}, and a second harmonic with frequency $\omega_{02} < \omega_{\text{best}}$, so that its effectiveness in inducing firing was relatively weak. The experiments showed that the ISIH's are multimodal, with basic period $T_{01} = 2\pi/\omega_{01}$ or $T_{02} = 2\pi/\omega_{02}$ according as the ratio $\eta = \gamma_{01}/\gamma_{02}$ of the amplitudes corresponding to ω_{01} and ω_{02} is relatively large or small. We noted earlier that for the bistable model a similar behavior follows from the Melnikov necessary condition from chaos. Even though the Melnikov scale factor $S(\omega_{01})$ is larger than $S(\omega_{02})$, if the ratio η is sufficiently small the motion can be dominated by the harmonic with inefficient frequency. That this is the case is shown by the simulations of Fig. 12.10a,b, for which $\omega_{01}/\omega_{\text{best}} = 1.0$, $\omega_{02}/\omega_{\text{best}} = 0.06$, $\beta = 0.25$, and $\sigma = 0.0075$. For Fig. 12.10a, $\gamma_{01} = 0.01$ and $\gamma_{02} = 0.3$ ($\eta = 0.033$). Apart from the two peaks at $t/T_{02} \ll 1$, which are ascribed to multiple firings, the spikes are grouped around the period T_{02} and its integer multiples.

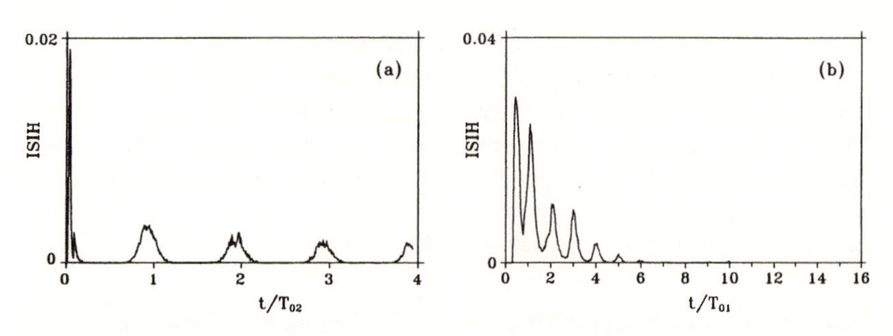

Figure 12.10. Interspike interval histograms simulated from Eq. 12.3.1 for excitation by two harmonics in the presence of noise. Dominant frequencies are controlled by (a) excitation with inefficient frequency higher than the "best" frequency and relatively large amplitude, (b) excitation with frequency close to "best" frequency.

For Fig. 12.10b, $\gamma_{01} = 0.1$ and $\gamma_{02} = 0.16$. Because the frequency ω_{01} is more effective than ω_{02} and the ratio η is not sufficiently small, period T_{01} controls.

12.5 CONCLUDING REMARKS

In this chapter it was shown that the qualitative properties of a bistable model to which the Melnikov approach is applicable are consistent with behavior observed in auditory nerve fiber experiments under a variety of types of excitation. It was noted that the Fitzhugh-Nagumo model's agreement with experiments appears to be unsatisfactory. Results of numerical simulations based on the bistable model confirmed the qualitative adequacy of the proposed model. Quantitative agreement depends upon the appropriate selection, by trial and error, of the shape of the system potential and the dissipation parameter. The behavior of the model is robust. Finally, we note that the proposed model is compatible with stochastic resonant behavior (see Chapter 8), whose role in the behavior of neuronal systems was noted by Collins, Imhoff, and Grigg 1996 and references quoted therein.

Appendix A1

Derivation of Expression for the Melnikov Function

As indicated in Section 2.4, we need to obtain the derivatives with respect to t of $\mathbf{x}_1^u(t; t_0, t')$ and $\mathbf{x}_1^s(t; t_0, t')$. We show how this is done for $\mathbf{x}_1^u(t; t_0, t')$. First, differentiation of Eq. 2.4.3a with respect to time, and the fact that $\dot{\mathbf{x}}_h = \mathbf{f}(\mathbf{x}_h)$ (see Eq. 2.3.1 in which $\epsilon = 0$), yield

$$\begin{aligned}\dot{\mathbf{x}}^u(t; t_0, t', \epsilon) &= \dot{\mathbf{x}}_h(t - t_0) + \epsilon\dot{\mathbf{x}}_1^u(t; t_0, t') + O(\epsilon^2)\\ &= \mathbf{f}[\mathbf{x}_h(t - t_0)] + \epsilon\dot{\mathbf{x}}_1^u(t; t_0, t') + O(\epsilon^2).\end{aligned} \qquad (A1.1)$$

Second, we use Eq. 2.3.1, in which we expand to first order the term $\mathbf{f}(\mathbf{x}^u)$ about the point with coordinate $\mathbf{x}_h(t - t_0)$, and then Eq. 2.4.3a, to obtain

$$\begin{aligned}\dot{\mathbf{x}}^u(t; t_0, t', \epsilon) &= \mathbf{f}[\mathbf{x}_h(t - t_0)] + (x_1^u - x_{h1})\partial\mathbf{f}[\mathbf{x}_h(t - t_0)]/\partial x_{h1}\\ &\quad + (x_2^u - x_{h2})\partial\mathbf{f}[\mathbf{x}_h(t - t_0)]/\partial x_{h2} + \epsilon\mathbf{g}(\mathbf{x}^u, t)\\ &= \mathbf{f}[\mathbf{x}_h(t - t_0)] + \epsilon x_{11}^u\partial\mathbf{f}[\mathbf{x}_h(t - t_0)]/\partial x_{h1}\\ &\quad + \epsilon x_{12}^u\partial\mathbf{f}[\dot{\mathbf{x}}_h(t - t_0)]/\partial x_{h2} + \epsilon\mathbf{g}(\mathbf{x}^u, t)\\ &\quad + O(\epsilon^2).\end{aligned} \qquad (A1.2)$$

From Eqs. A1.1 and A1.2,

$$\begin{aligned}\dot{\mathbf{x}}_1^u &= x_{11}^u\partial\mathbf{f}[\mathbf{x}_h(t - t_0)]/\partial x_{h1} + x_{12}^u\partial\mathbf{f}[\mathbf{x}_h(t - t_0)]/\partial x_{h2}\\ &\quad + \mathbf{g}(\mathbf{x}^u, t) + O(\epsilon).\end{aligned} \qquad (A1.3)$$

A similar expression holds for $\dot{\mathbf{x}}_1^u$. We now differentiate Eq. 2.4.5a and then use Eq. A1.3:

$$\begin{aligned}\dot{\Delta}_\epsilon^u(t; t_0, t') &= \dot{\mathbf{f}}[\mathbf{x}_h(t - t_0)] \wedge \epsilon\mathbf{x}_1^u + \mathbf{f}[\mathbf{x}_h(t - t_0)] \wedge \epsilon\dot{\mathbf{x}}_1^u\\ &= \{\partial\mathbf{f}[\mathbf{x}_h(t - t_0)]/\partial x_{h1}\dot{x}_{h1}\end{aligned}$$

$$+\partial \mathbf{f}[\mathbf{x}_h(t-t_0)]/\partial x_{h2}\dot{x}_{h2}\} \wedge \epsilon \mathbf{x}_1^u$$
$$+\mathbf{f}[\mathbf{x}_h(t-t_0)] \wedge \epsilon\{x_{11}^u \partial \mathbf{f}[\mathbf{x}_h(t-t_0)]/\partial x_{h1}$$
$$+x_{12}^u \partial \mathbf{f}[\mathbf{x}_h(t-t_0)]/\partial x_{h2} + \mathbf{g}(\mathbf{x}^u, t)\}$$
$$+O(\epsilon^2). \tag{A1.4}$$

Since $\mathbf{f} = f_1\mathbf{i}_1 + f_2\mathbf{i}_2$, $\dot{\mathbf{x}}_h = \mathbf{f}(\mathbf{x}_h)$, $\mathbf{f}$ is Hamiltonian, and $\mathbf{i}_j \wedge \mathbf{i}_j = 0$, after some algebra we obtain

$$\dot{\Delta}_\epsilon^u(t; t_0, t') = \epsilon \mathbf{f}[\mathbf{x}_h(t-t_0)] \wedge \mathbf{g}(\mathbf{x}^u, t) + O(\epsilon^2), \tag{A1.5a}$$

or, using Eq. 2.4.3a,

$$\dot{\Delta}_\epsilon^u(t; t_0, t') = \epsilon \mathbf{f}[\mathbf{x}_h(t-t_0)] \wedge \mathbf{g}[\mathbf{x}_h(t-t_0), t] + O(\epsilon^2). \tag{A1.5b}$$

Integration of Eq. A1.5 with respect to time from $-\infty$ to t' yields

$$\Delta_\epsilon^u(t; t_0, t') = \epsilon \int_{-\infty}^{t'} \mathbf{f}[\mathbf{x}_h(t-t_0)] \wedge \mathbf{g}[\mathbf{x}_h, (t-t_0), t]\, dt + O(\epsilon^2), \tag{A1.6}$$

where we used the fact that $\Delta_\epsilon^u(-\infty; t_0, t') = 0$ (since, for the fixed point $\mathbf{x}_h(-\infty) = 0$, by Eq. 2.1.1.a $\mathbf{f}(0) = 0$). Similarly

$$\Delta_\epsilon^s(t; t_0, t') = \epsilon \int_{t'}^{\infty} \mathbf{f}[\mathbf{x}_h(t-t_0)] \wedge \mathbf{g}[\mathbf{x}_h(t-t_0), t]\, dt + O(\epsilon^2). \tag{A1.7}$$

With the change of variable $\zeta = t - t_0$, it follows from Eqs. 2.4.5, A1.6, and A1.7 that

$$\Delta_\epsilon(t'; t_0, t') = \epsilon \int_{-\infty}^{\infty} \{\mathbf{f}[\mathbf{x}_h(\zeta)] \wedge \mathbf{g}[\mathbf{x}_h(\zeta), \zeta + t_0]\}\, d\zeta + O(\epsilon^2), \tag{A1.8}$$

so that, in view of Eq. 2.4.1, Eq. 2.4.6 holds.

Note that, for any fixed plane $t' =$ const, a change of the variable t_0 corresponds to a change of the position of the point P' along a homoclinic orbit in that plane (Figs. 2.3 and 2.5).

Appendix A2

Construction of Phase Space Slice through Stable and Unstable Manifolds

We describe the construction of the intersections with a plane of section (i.e., a phase space slice) of the invariant manifolds of the system defined by the equations of motion

$$\frac{dx_1}{dt} = x_2, \tag{A2.1a}$$

$$\frac{dx_2}{dt} = -V'(x_1) + \epsilon[f(x_1, x_2) + g(t - t_1)]. \tag{A2.1b}$$

We also refer to Eqs. A2.1 as the system's equations of motion in forward time, that is, the equations describing the evolution of the system as the time t increases. The equations of motion in reverse time—the equations describing the evolution of the system as the time t decreases—are obtained from Eqs. A2.1 by reversing the orientation of the time axis, that is, by the transformation $t = -\tau$, $t_1 = -\tau_1$. The result is

$$\frac{dx_1}{d\tau} = -x_2, \tag{A2.2a}$$

$$\frac{dx_2}{d\tau} = -\{-V'(x_1) + \epsilon[f(x_1, x_2) + g(-(\tau - \tau_1))]\}. \tag{A2.2b}$$

where $x_i \equiv x_i(-\tau) = x_i(t)$ $(i = 1, 2)$.

As an exercise, the reader may check this result in a specific case by performing two integrations. First, the equation of motion in forward time is integrated to time $t = t_f$, the initial conditions being $[x_1(0), x_2(0)]$ at $t = 0$. The result of the integration is $[x_1(t_f), x_2(t_f)]$. Second, the reverse equation of motion is integrated to time $\tau = 0$, the initial conditions being $[x_1(-\tau_f), x_2(-\tau_f)]$ at time $\tau = -\tau_f = t_f$. The result of the integration should be $[x_1(0), x_2(0)]$.

A2.1 PROCEDURE FOR CONSTRUCTING STABLE AND UNSTABLE MANIFOLDS

The construction entails the following steps:

1. Choose the coordinate t_s of the phase space slice (plane of section) whose intersection with the perturbed stable and unstable manifolds is sought.
2. For sufficiently small ϵ the set Γ of homoclinic points of the unperturbed system (Eqs. A2.1 in which $\epsilon = 0$) persists, that is, it is mapped onto a curve Γ_ϵ (Fig. 2.4). The second step consists of finding the points O_{1-} and O_{1+} where Γ_ϵ intersects the two planes of section defined by the coordinates $t_s - t_d$ and $t_s + t_d$ ($t_d > 0$) (Fig. A2.1).
3. Obtain the eigenvalues and eigenvectors of the linearized equations of motion at points O_{1-} and O_{1+}.
4. Using as initial conditions the coordinates of a point close to O_{1-} on the linearized system's unstable eigenvectors (i.e., a point belonging to the system's local unstable manifolds), integrate the forward equations of motion from time $t_s - t_d$ to time t_s. The solution at time t_s belongs to the global unstable manifold being constructed. Repeat the procedure

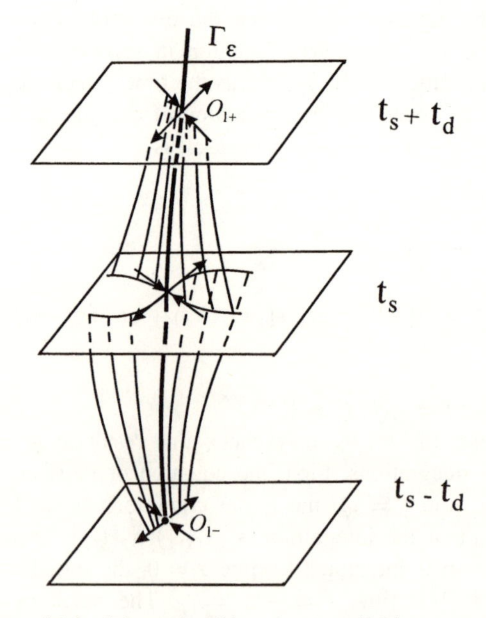

Figure A2.1. Schematic of invariant manifold construction.

using as initial conditions a sufficient number of points close to O_{1-} on the unstable eigenvectors (Fig. A2.1).

5. Using as initial conditions the coordinates of a point close to O_{1+} on the linearized system's stable eigenvectors, integrate the reverse equations of motion from time $-(\tau_s + \tau_d) = t_s + t_d$ to time $-\tau_s = t_s$. The solution at time $-\tau_s = t_s$ belongs to the stable manifold being constructed. Repeat the procedure using as initial conditions a sufficient number of points close to O_{1+} on the stable eigenvectors (Fig. A2.1).

A2.2 EIGENVALUES AND EIGENVECTORS OF LINEARIZED SYSTEM

We consider for specificity the standard Duffing–Holmes oscillator, that is,

$$V'(x_1) = x_1 - x_1^3, f(x_1, x_2) = \beta x_2, \quad \text{(A2.3a,b)}$$

$\beta > 0$. The variational equation corresponding to Eq. A2.1 is (Section 2.1.1)

$$\delta\dot{x}_1 = \delta\dot{x}_2, \quad \text{(A2.4a)}$$

$$\delta\dot{x}_2 = (1 - 3x_1^2)\delta x_1 - \beta\delta x_2. \quad \text{(A2.4b)}$$

The solutions of Eqs. A2.4 are $\delta x_1 = u_{1,s}\exp(\lambda_s t)$, $\delta x_2 = u_{2,s}\exp(\lambda_s t)$ and $\delta x_1 = u_{1,u}\exp(\lambda_u t)$, $\delta x_2 = u_{2,u}\exp(\lambda_u t)$. The eigenvalues λ_s, λ_u at point $\{x_1, x_2\}$ are obtained from the characteristic equation $|a_{ij} - I\lambda| = 0$, where $[a_{ij}]$ $(i, j = 1, 2)$ is the matrix of the coefficients of $\delta x_1, \delta x_2$ in Eqs. A2.4 $(a_{11} = 0, a_{12} = 1, a_{21} = 1 - 3x_1^2, a_{22} = -\beta)$, and I is the unit matrix (see Example 2.1.1), that is,

$$\begin{vmatrix} -\lambda & 1 \\ 1 - 3x_1^2 & -\beta - \lambda \end{vmatrix} = 0. \quad \text{(A2.5)}$$

Equation A2.5 yields the solutions

$$\lambda_{u,s} = \tfrac{1}{2}[-\beta \pm (\beta^2 + 4(1 - 3x_1^2)^{1/2})]. \quad \text{(A2.6)}$$

For small values of x_1 and β, λ_u is positive and somewhat smaller than 1, and λ_s is negative and somewhat smaller than -1.

Substituting the exponential solutions in Eqs. A2.4 we get

$$a_{11}u_{1,s} + a_{12}u_{2,s} = \lambda_s u_{1,s}, \quad \text{(A2.7a)}$$

$$a_{21}u_{1,s} + a_{22}u_{2,s} = \lambda_s u_{2,s} \quad \text{(A2.7b)}$$

with a similar set of equations for λ_u. Using the first equation in each set and recalling that $a_{11} = 0$, $a_{12} = 1$, we obtain the slopes of the corresponding eigenvectors:

$$u_{2,s}/u_{1,s} = \lambda_s, \tag{A2.8a}$$

$$u_{2,u}/u_{1,u} = \lambda_u. \tag{A2.8b}$$

Denote by $x_{1,\epsilon}$, $x_{2,\epsilon}$ the coordinates of the intersection of the curve Γ_ϵ with the plane of section being considered. The equations of the local stable and unstable manifolds in that plane are

$$x^s_{2,\,\mathrm{loc}} = x_{2,\epsilon} + \lambda_s(x^s_{1,\,\mathrm{loc}} - x_{1,\epsilon}), \tag{A2.9a}$$

$$x^u_{2,\,\mathrm{loc}} = x_{s2,\epsilon} + \lambda_u(x^u_{1,\,\mathrm{loc}} - x_{1,\epsilon}). \tag{A2.9b}$$

A2.3 OBTAINING THE INTERSECTIONS O_{1-} AND O_{1+} OF CURVE Γ_ϵ, WITH PLANES OF SECTION $t = t_s - t_d$ AND $t = t_s + t_d$

We describe a method for obtaining the intersections by successive approximations. For specificity we consider point O_{1-}. Construct a quadrangle believed to enclose the point O_{1-} being sought (Fig. A2.2). The slopes of the quadrangle sides originating at points A and C should be equal to the values

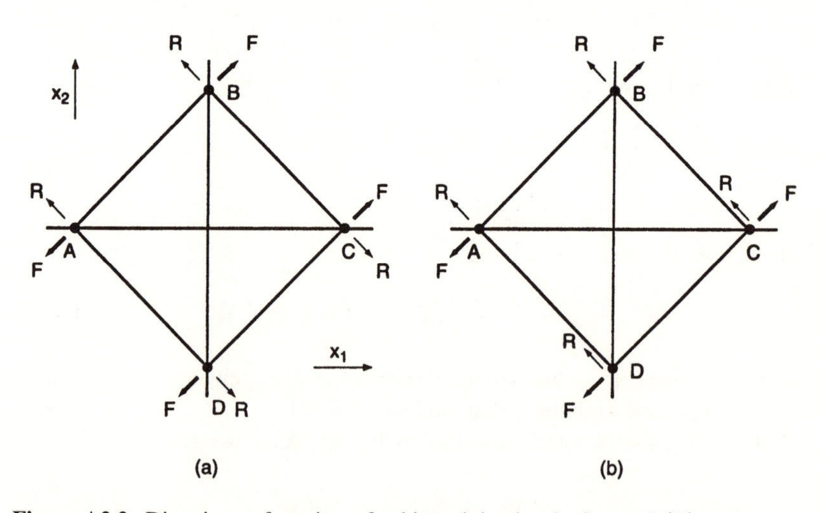

Figure A2.2. Directions of motion of orbits originating in forward (F) and reverse (R) time at corners of quadrangle (a) enclosing and (b) not enclosing point O_{1-}.

of λ_s and λ_u at those points. Integrate the forward and the reverse equations of motion using the coordinates of points A, B, C, and D as initial conditions. If the directions of motion are as in Fig. A2.2a, then O_{1-} is inside the quadrangle. If the directions are as in Fig. A2.2b, say, then the quadrangle does not enclose O_{1-}. The procedure is repeated with quadrangles A', B', C', D', $A'', B'', C'', D'', \ldots,$ contained in $A, B, C, D, A', B', C', D', \ldots,$ respectively, until the quadrangle containing the point O_{1-} is sufficiently small that the approximation to which O_{1-} is obtained is satisfactory. Once the coordinates of O_{1-} are available, the curve Γ_ϵ and its intersections with the planes of interest are obtained by integrating the equation of motion in forward time with those coordinates as initial conditions.

A2.4 NUMERICAL EXAMPLE

For the standard Duffing–Holmes oscillator, we assume $\epsilon\beta = 0.1, t_1 = 0$, and a forcing consisting of a harmonic oscillation and a sum of harmonics that may be viewed as an approximate representation of a Gaussian process path (Simiu and Frey, 1993a):

$$\epsilon g(t) = \epsilon\gamma \cos(\omega t) + \epsilon\sigma(2/N) \sum_{k=1}^{N} \cos(\omega_k t + \phi_k), \qquad \text{(A2.10)}$$

where $\omega = 1$ and $\epsilon\gamma$ is chosen so that, for $\sigma = 0$, the Melnikov function has double zeros, that is, $-4\beta/3 + \gamma S(\omega) = 0$, $S(\omega) = \sqrt{2}\pi\omega \operatorname{sech}(\pi\omega/2)$ being the Melnikov scale factor corresponding to $\gamma = 1$ in Eq. 2.5.17. This condition yields $\epsilon\gamma = 0.07530181$. For $\sigma \neq 0$ the Melnikov function will have simple zeros. We assume $N = 15$, $\epsilon\sigma = 0.02$ (i.e., the common amplitude of the terms in the sum of Eq. A2.10 is $0.02(2/15)^{1/2} = 0.00730297$), and the following frequencies and phase angles:

$$[\omega_1, \omega_2, \ldots, \omega_{15}] = [0.2177, 0.6147, 0.9834, 1.3966, 1.8073, 2.1843, 2.6103, 2.8328, 3.4128, 3.8018, 4.1888, 4.5886, 5.006, 5.3853, 6.1843],$$

$$[\phi_1, \phi_2, \ldots, \phi_{15}] = [3.0473, 2.5509, 5.0328, 3.9521, 0.7979, 3.1792, 6.1952, 3.4808, 2.5195, 3.3489, 1.6888, 3.3552, 1.7216, 3.7384, 2.0985].$$

We construct the intersection of the stable and unstable manifolds at time $t = t_s = 10\pi$. We choose $t_d = 10\pi$. For $t = t_s - t_d = 0$, the coordinates of point O_{1-} are found to be $[-0.024207850724634, 0.002384710615867]$.

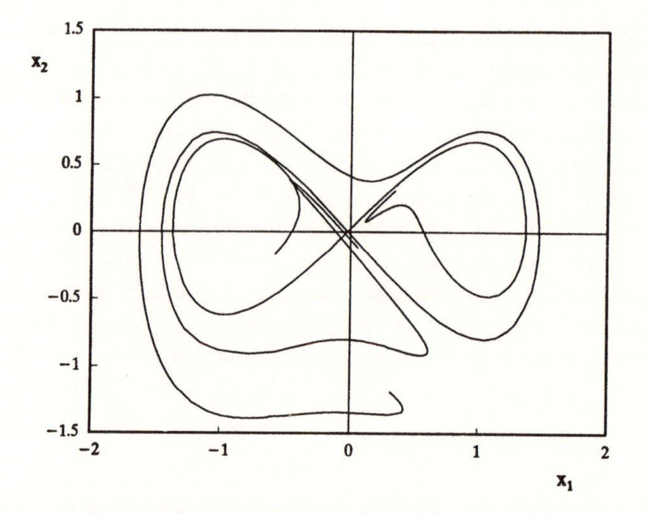

Figure A2.3. Phase space slice for the Duffing–Holmes equation excited by a harmonic function and an approximate representation of a Gaussian noise path (after Simiu and Frey, 1993a).

The corresponding eigenvalues are $\lambda_u = 0.950370901161105$, $\lambda_s = -1.050307901161105$. The phase space slice at $t = 10\pi$ is shown in Fig. A2.3. An intersection between the stable and unstable manifolds is seen to occur in the second quadrant.

Appendix A3

Topological Conjugacy

Maps q_1: $A \to A$ and q_2: $B \to B$ are *topologically conjugate* if there exists a homeomorphism h (i.e., if there exists a one-to-one, onto (see footnote 1, Section 3.3), and continuous function h with a continuous inverse) such that

$$h \circ q_1(\mathbf{x}) = q_2 \circ h(\mathbf{x}). \tag{A3.1}$$

Topologically conjugate maps have completely equivalent dynamics. We show that this is the case in the following example.

Example

Assume $h = \frac{1}{2}(1 - x)$, $q_1 = 2x^2 - 1$, and let $q_2 = 4x(1 - x)$ be defined on the interval $[0, 1]$. It is easily seen that h is one to one, onto, and continuous, and has a continuous inverse, that is, that h is a homeomorphism. The reader can verify immediately that $h \circ q_1(\mathbf{x}) = \frac{1}{2}[1 - (2x^2 - 1)] = 1 - x^2 = q_2 \circ\ h(\mathbf{x})$. The maps q_1 and q_2 are therefore topologically conjugate. The fixed points of the map q_1 are obtained from the condition that the iterate of x under q_1 is equal to x, that is, $2x^2 - 1 = x$, or $x_1 = 1$, $x_2 = -\frac{1}{2}$. These points are mapped by h to $h(1) = 0$ and $h(-\frac{1}{2}) = \frac{3}{4}$. From the condition $4x(1 - x) = x$ it follows that the fixed points of the map q_2 are indeed 0 and $\frac{3}{4}$.

Since h has a continuous inverse, and $h^{-1} \circ h \circ q_1(\mathbf{x}) \equiv q_1(\mathbf{x})$, it follows from Eq. A3.1

$$q_1(\mathbf{x}) = h^{-1} \circ q_2 \circ h(\mathbf{x}). \tag{A3.2}$$

The orbit of $\mathbf{x}$ under q_1 in A is

$$\{\dots q_1^{-n}(\mathbf{x}), \dots, q_1^{-1}(\mathbf{x}), \mathbf{x}, q_1(\mathbf{x}), \dots, q_1^{n}(\mathbf{x}), \dots, \}. \tag{A3.3}$$

From Eq. A3.2 we obtain the following results:

$$\begin{aligned} q_1^n(\mathbf{x}) &= (h^{-1} \circ q_2 \circ h) \circ (h^{-1} \circ q_2 \circ h) \circ \cdots \circ (h^{-1} \circ q_2 \circ h(\mathbf{x})) \\ &= h^{-1} \circ q_2^n \circ h(\mathbf{x}), \qquad n \geq 0, \end{aligned} \tag{A3.4}$$

$$q_1^{-n} = h^{-1} \circ q_2^{-n} \circ h(\mathbf{x}), \qquad n \geq 0. \tag{A3.5}$$

From Eqs. A3.4 and A3.5 we get

$$h \circ q_1^n(\mathbf{x}) = q_2^n \circ h(\mathbf{x}) \text{ for all } n, \tag{A3.6}$$

that is, the orbit of $\mathbf{x}$ under q_1 is mapped by h to the orbit of $h(\mathbf{x})$ under q_2. It follows that if $\mathbf{x}_0$ is a fixed point under q_1, then $h(\mathbf{x}_0)$ is a fixed point under q_2.

Appendix A4

Properties of Space Σ_2

The space Σ_2 has the following properties:

1. *Σ_2 contains a countable infinity of periodic orbits.* This property is discussed in Section 3.4.
2. *Periodic points form a dense subset of Σ_2.*[1] To show this, we consider an arbitrary point in Σ_2,

$$s = \{\cdots s_{-n-2}s_{-n-1}s_{-n} \cdots s_{-2}s_{-1} \cdot s_0 s_1 s_2 \cdots s_n s_{n+1} s_{n+2} \cdots\},$$

   and the sequence of periodic points of period n,

$$\tau_n = \{\cdots t_{-n} \cdots t_{-2}t_{-1} \cdot t_0 t_1 t_2 \cdots t_n \quad t_{-n} \cdots t_{-2}t_{-1}t_0 t_1 t_2 \cdots t_n \cdots\},$$

   whose repeating subsequences are such that $t_i = s_i$ for $i = 0, \pm 1, \ldots, \pm n$. The distance between τ_n and s is $d(s, \tau_n) \leq 1/2^{n-1}$, so τ_n converges to s, that is, s is a limit point of τ_n. The set Σ_2 consists of the union of all periodic points and their limit points.
3. *Σ_2 contains an uncountable number of nonperiodic orbits.* Each nonperiodic orbit is associated with a nonrepeating sequence. The fact that the number of such sequences is uncountable is proved by Wiggins (1990, p. 431).
4. *Σ_2 contains a dense orbit under iteration by σ.* Dense orbits visit the neighborhood of any given point in Σ_2. Maps with this property are called *topologically transitive*. A sequence s_d associated with a dense orbit can be constructed as follows:

$$s_d = \{\text{Block 1 Block 2 Block 3} \ldots\}$$

[1]A subset Σ_p of a set Σ is *dense* in Σ if Σ consists of all points in Σ_p together with their *limit points* (also called accumulation points). A point $\mathbf{x} \in \Sigma$ is a limit point of Σ_p if there is a sequence of points $\mathbf{x} \in \Sigma_p$ that converges to $\mathbf{x}$.

where Block i consists of all subsequences of length i. For example, Block 2 consists of the subsequences 00 01 10 11. Some point obtained from the sequence s_d under iteration by σ will be arbitrarily close to any point p of Σ_2, that is, p and some iterate of s_d will have an arbitrarily large number of the same entries on both sides of the decimal point.

Appendix A5

Elements of Probability Theory

This Appendix reviews

- the definition of probabilities in terms of relative frequencies
- the addition and multiplication rules of probability theory
- the definition of probability distributions
- the definition of simple descriptors of stochastic variable behavior, including the mean and the variance
- the definitions of the Poisson distribution, the normal (Gaussian) distribution, and non-Gaussian distributions

A5.1 RELATIVE FREQUENCY AND PROBABILITY OF AN EVENT. RANDOMNESS

Consider an experiment whose only possible outcomes are the occurrence or nonoccurrence of an event A. The probability of occurrence $P(A)$ of event A is the limit for large n of the *relative frequency* m/n, where m is the total number of occurrences of the event A in n trials of the experiment. It is assumed that a sequence (S) of trials satisfies the condition of randomness, meaning that the relative frequency of A must have the same limiting value in that sequence as in any partial sequences that might be selected from it in any arbitrary way, the number of terms in the partial sequences being sufficiently large, and the selection being made in the absence of any information on the outcomes of the experiment (von Mises, 1957).

A5.2 FUNDAMENTAL RELATIONS

A5.2.1 Addition of Probabilities

Consider two mutually exclusive events A_1 and A_2 associated with an experiment. (Mutually exclusive events are events that cannot occur in the same experiment: for example, a head and a tail in coin toss, or a two and a six in a throw of a die.) The event that either A_1 or A_2 will occur is denoted by $A_1 \cup A_2$. Its probability is

$$P(A_1 \cup A_2) = P(A_1) + P(A_2). \tag{A5.1}$$

The empirical basis of Eq. A5.1 lies in the fact that, if the relative frequency of event A_1 is m_1/n and that of A_2 is m_2/n, then the relative frequency of occurrence of either A_1 or A_2 is $(m_1 + m_2)/n$.

Example

In a fair die the probability of a two is 1/6 and the probability of a six is 1/6. By virtue of Eq. A5.1, the probability of either a two or a six is 1/3.

A5.2.2 Compound and Conditional Probabilities. The Multiplication Rule

Consider two events A and B that may occur at the same time. The probability of the event that A and B will occur simultaneously is called the *compound probability* of events A and B and is denoted *by* $P(A \cap B)$. The probability of a given event A given that event B has already occurred is referred to as the *conditional probability* of event A under the condition that event B has occurred. It is denoted by $P(A|B)$, and is defined as follows:

$$P(A|B) = \frac{P(A \cap B)}{P(B)}, \tag{A5.2}$$

where $P(B)$ is assumed to be different from zero.

Example

In a certain region records show that, on average, there are in a year 100 cold days (event A), 200 sunny days (event B), and 30 days that are both cold and sunny (event $A \cap B$). The probability $P(A|B)$ that a day is cold under the condition that the day is sunny is $P(A|B) = P(A \cap B)/P(B) = (30/365)/(200/365) = 30/200$.

From Eq. A5.2 it follows that

$$P(A \cap B) = P(A|B)P(B). \tag{A5.3}$$

Equation A5.3 is called the *multiplication rule* of probability theory.

A5.2.3 Independence

Two events A and B for which

$$P(A|B) = P(A) \tag{A5.4}$$

are called *stochastically independent*. From Eqs. A5.3 and A5.4 it follows that an alternative definition of independence is

$$P(A \cap B) = P(A)P(B). \tag{A5.5}$$

A5.3 STOCHASTIC VARIABLES. HISTOGRAMS AND PROBABILITY DISTRIBUTIONS

A5.3.1 Definition of Stochastic Variables. Continuous and Discrete Stochastic Variables

Let a numerical value be assigned to each of the events that may occur as a result of an experiment. The resulting set of numbers is defined as a stochastic variable. Stochastic variables are called discrete or continuous according as they take on values restricted to a finite set of numbers (e.g., the set of numbers 0 and 1 assigned to the outcomes heads and tails, respectively, in a coin-toss experiment), or any value on a segment of the real axis (e.g., the air temperature at a given location). It is customary to denote stochastic variables by capital Roman letters, for example, X, Y, Z. Specific values that may be taken on by these variables are then denoted by the corresponding lower-case letters x, y, z.

A5.3.2 Histograms, Probability Density Functions, Cumulative Distribution Functions

Partition the range of a continuous stochastic variable X associated with an experiment into equal intervals of width ΔX. If the experiment is carried out n times, the number of times X has taken on values $X_0 < X \leq X_1$, $X_1 < X \leq X_2, \ldots, X_{m-1} < X \leq X_m$, where $X_i - X_{i-1} = \Delta X (i = 1, 2, \ldots, m)$, is $n_1, n_2, \ldots, n_m$, respectively, and $n_1 + n_2 + \cdots + n_m = n$. A diagram in which the numbers $n_1, n_2, \ldots, n_m$ are plotted as functions of the respective intervals is called a *histogram*.

If we divide the ordinates $n_1, n_2, \ldots, n_m$ of a histogram by $n\Delta X$ we obtain a diagram called the *frequency density distribution*. The total area under a frequency distribution diagram is

$$(n_1 + n_2 + \cdots + n_m)\Delta X/(n\Delta X) = 1.$$

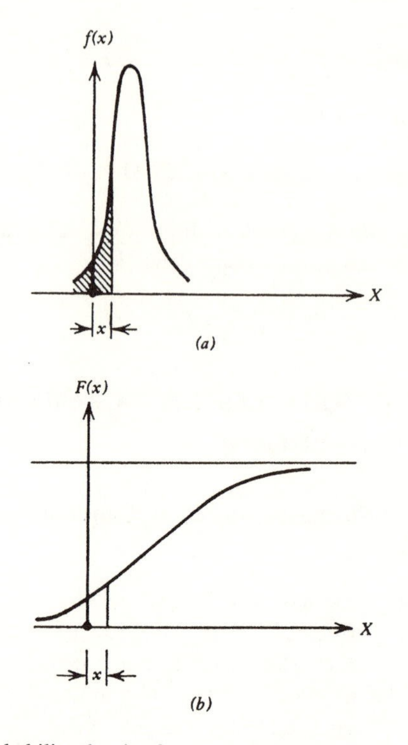

Figure A5.1. (a) Probability density function of a stochastic variable X; (b) cumulative distribution function of X.

As ΔX becomes very small and as n becomes very large, the ordinates of the frequency density distribution approach values denoted by $f(x)$, where x denotes a value that may be taken on by the variable X. The function $f(x)$ is called the *probability density function* of the variable X (Fig. 4.1a). The probability of the event $x < X \leq x + dx$ is equal to $f(x)\,dx$, and

$$\int_{-\infty}^{\infty} f(x)\,dx = 1. \tag{A5.6}$$

The probability that $X \leq x$ is the *cumulative distribution function* of the variable X, and can be written as

$$F(x) = \int_{-\infty}^{x} f(x)\,dx \tag{A5.7}$$

(Fig. A5.1b). By Eq. A5.7, the ordinate at $X = x$ in Fig. A5.1b is equal to the shaded area of Fig. A5.1a. As shown by Eqs. A5.6 and A5.7, $F(x)$ approaches unity for very large x.

A5.3.3 Joint Probability Distributions

Let X and Y be two continuous stochastic variables and let $f_{xy}(x, y)\,dx\,dy$ denote the probability that $x < X \leq x + dx$ and $y < Y \leq y + dy$. (The subscript is used for the sake of clarity but is not obligatory.) The function $f_{xy}(x, y)$ is then called the *joint probability density function* of the stochastic variables X, Y (Fig. A5.2). The probability that $X \leq x$ and $Y \leq y$ is called the *joint cumulative probability distribution* of X and Y and is denoted by $F_{xy}(x, y)$. From the definition of $f_{xy}(x, y)\,dx\,dy$ it follows that

$$F_{xy}(x, y) = \int_{-\infty}^{x} \int_{-\infty}^{y} f_{xy}(x, y) dx\, dy. \tag{A5.8}$$

If $f_{xy}(x, y)$ is known, the probability that $x < X \leq x + dx$, denoted by $f_x(x)\,dx$, is obtained by applying the addition rule to the probabilities $f_{xy}(x, y)\,dx\,dy$ over the entire Y domain, that is,

$$f_x(x) = \int_{-\infty}^{\infty} f_{xy}(x, y)\, dy. \tag{A5.9}$$

The function $f_y(x)$ is called the *marginal probability density function* of X.

The probability that $y < Y \leq y + dy$ under the condition that $x = X$ is denoted by $f(y|x)\,dy$. The function $f(y|x)$ is called the *conditional proba-*

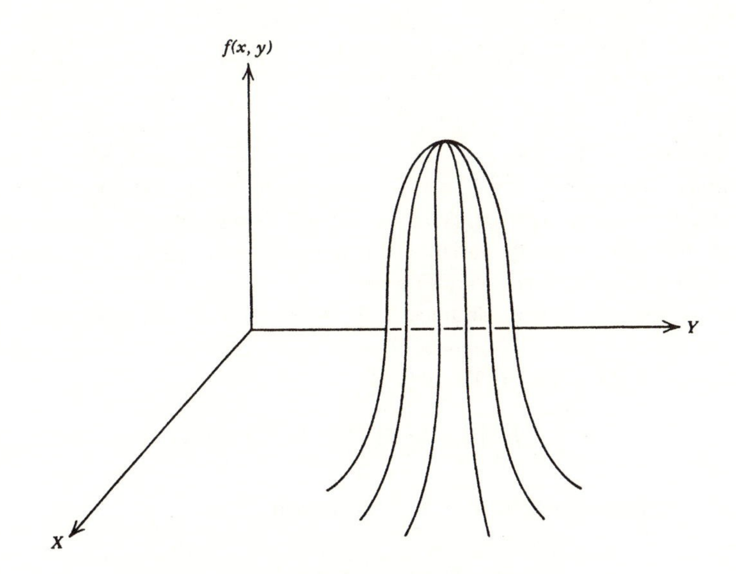

Figure A5.2. Probability density function $f_{xy}(x, y)$.

bility density function of y given that $X = x$. If Eq. A5.2 is used, it follows that

$$f(y|x) = \frac{f_{xy}(x, y)}{f_x(x)}. \tag{A5.10}$$

If X and Y are independent, $f(y|x) = f_y(y)$, and

$$f_{xy}(x, y) = f_x(x) f_y(y). \tag{A5.11}$$

Similar definitions apply to any number of continuous stochastic variables.

A5.4 DESCRIPTORS OF STOCHASTIC VARIABLE BEHAVIOR

The full description of the probabilistic behavior of a stochastic variable is provided by its probability distribution (for several variables, by their joint probability distribution). For distributions of interest in this text, useful if less detailed information is provided for one variable X by the *mean value* (or *expected value*, or *expectation*)

$$E(X) = \int_{-\infty}^{\infty} x f(x)\, dx \tag{A5.12a}$$

and the first nth-order *moments* about the mean

$$E\{[X - E(X)]^n\} = \int_{-\infty}^{\infty} [x - E(X)]^n f(x)\, dx \tag{A5.12b}$$

where n is finite. In Eqs. A5.12 and subsequently in this Appendix we omit for convenience the subscripts in the notation for the probability density functions. The second-, third-, and fourth-order moments about the mean are called the *variance*, *skewness*, and *kurtosis* of X, respectively. The *standard deviation* of the stochastic variable X is denoted by σ_x and is defined as the square root of its variance $\mathrm{Var}(X) \equiv \sigma_x^2$.

The *covariance* of two variables X, Y, denoted by σ_{xy}, is defined as

$$E\{[X - E(X)][Y - E(Y)]\} = \int_{-\infty}^{\infty}\int_{-\infty}^{\infty} [x - E(X)][y - E(Y)] f(x, y)\, dx\, dy. \tag{A5.12c}$$

The *correlation coefficient* of X, Y is defined as

$$\rho = \sigma_{xy}/(\sigma_x \sigma_y). \tag{A5.12d}$$

Equations A5.12a–c follow from the relation between the probability density function and relative frequencies. For example, consider a discrete variable X. Its mean is

$$E(X) = \sum_{i=1}^{m} x_i \frac{n_i}{n} \tag{A5.12e}$$

$$= \sum_{i=1}^{m} x_i f_i \tag{A5.12f}$$

where $f_i = n_i/n(i = 1, 2, \ldots, m)$ is the relative frequency of the value x_i taken on by X. Equation A5.12a is the counterpart for the case of a continuous variable of Eq. A5.12f. A similar argument holds for moments of order n or for the covariance of X and Y.

A5.5 THE POISSON DISTRIBUTION

The Poisson distribution, defined in this section, is used in Chapter 5 for the calculation of probabilities of occurrence of rare events where a threshold is exceeded by a stochastic process. Consider a stationary process representing the number of events occurring during the time interval $[0, \tau)$. Assume that (1) the numbers of events occurring in nonoverlapping time intervals are mutually independent, and (2) in a small time interval ξ the probability of occurrence of one event is $\lambda\xi + o(\xi)$ and that of multiple events is $o(\xi)$, where $o(\xi)/\xi \to 0$ as $\xi \to 0$, and $\lambda > 0$ is a fixed number referred to as the average rate of arrival of the process. It can be shown that the probability of exactly n events occurring during the time interval τ is

$$p(n, \tau) = \frac{(\lambda\tau)^n}{n!} e^{-\lambda\tau} \quad (n = 0, 1, 2, 3, \ldots,). \tag{A5.13}$$

Substitution of Eq. A5.13 into Eqs. A5.12a and A5.12b ($n = 2$) yields the result that both the expected value and the variance of n are equal to $\lambda\tau$.

A5.6 THE NORMAL DISTRIBUTION AND ITS PHYSICAL SIGNIFICANCE. NON-GAUSSIAN DISTRIBUTIONS

Consider a stochastic variable X that is the sum of n stochastic, independent contributions of similar order of magnitude. It can be proved that, under certain general conditions, in the limit $n \to \infty$ the probability density function of X is

$$f(x) = \frac{1}{(2\pi)^{1/2}\sigma_x} \exp\left[\frac{-(x - \mu_x)^2}{2\sigma_x^2}\right], \tag{A5.14}$$

where $\mu_x = E(X)$ and $\sigma_x^2 = \mathrm{Var}(X)$ are the mean value and the variance of X, respectively. This statement is known as the *central limit theorem.* The distribution (A5.14) is called *normal* or *Gaussian.* As shown by Eq. A5.14, the Gaussian distribution is fully determined if the mean and standard deviation of the variable are known. Note that the *tails* of the distribution are infinitely long, meaning that $f(x)$ has finite albeit small positive values for arbitrarily large values $|x|$.

In physically realizable systems the number n of contributions is of necessity finite, although it can be very large. Distributions applicable to such systems, which for mathematical convenience are commonly idealized as Gaussian, are therefore only approximately Gaussian, and their tails, although possibly very long, are finite, rather than infinite. It is shown in Chapter 5 that this observation is relevant to the development of the stochastic Melnikov approach as applicable to physically realizable systems.

The *joint Gaussian distribution* of two variables X and Y with zero means is

$$f(x,y) = \frac{1}{2\pi\sigma_x\sigma_y(1-\rho^2)^{1/2}} \times \exp\left[-\frac{1}{2(1-\rho^2)}(x^2/\sigma_x^2 - 2\rho xy/(\sigma_x\sigma_y) + x^2/\sigma_y^2)\right], \quad \text{(A5.15)}$$

where the correlation coefficient ρ is defined by Eq. A5.12d.

All distributions that are not Gaussian are referred to as *non-Gaussian.*

Appendix A6

Mean Upcrossing Rate τ_u^{-1} for Gaussian Processes

In the subsequent derivations we follow Rice (1954). We consider a differentiable stationary Gaussian process $x(t)$ with zero mean, and denote by $f_{x\dot{x}}(x, \dot{x}; t_1)$ the joint probability density function of x and $\dot{x}$ at time t_1. The process $x(t)$ upcrosses (i.e., crosses with positive slope) the level k in $(t_1, t_1 + dt)$ if $x(t_1) < k$ and $x(t_1 + dt) > k$. It can be shown that the average number of upcrossings of $x(t)$ in $(t_1, t_1 + dt)$ is

$$dt \int_0^\infty k f_{x\dot{x}}(x, \dot{x}; t_1)\, d\dot{x}. \tag{A6.1}$$

It is a classic result that a linear combination of Gaussian processes is Gaussian. The derivative $\dot{x}(t)$ of the Gaussian process $x(t)$ is proportional to the difference between two Gaussian processes, $x(t)$ and $x(t + dt)$, and is therefore also Gaussian. Since both $x(t)$ and $\dot{x}(t)$ are Gaussian, so is their joint distribution $f_{x\dot{x}}$. To perform the integration in Eq. A6.1 we may therefore make use of the expression for the joint Gaussian distribution, Eq. A5.15. We now seek the correlation coefficient in that expression.

Let us consider the Bennett-Rice noise approximation of the process $x(t)$:

$$x(t) = \sum_{k=1}^{N} a_k \cos(\omega_k t + \phi_{0k}) \tag{A6.2}$$

(Eq. 4.2.5), where $a_k = [2\Psi_x(\omega_k)\Delta\omega/(2\pi)]^{1/2}$, ϕ_{0k} are independent and uniformly distributed over the interval $[0, 2\pi]$, $\omega_k = k\Delta\omega$, $\Delta\omega = \omega_{\text{cut}}/N$, and ω_{cut} is the frequency beyond which the spectrum $\Psi_x(\omega)$ vanishes or becomes negligible (the cutoff frequency). As was indicated in Section 4.2.2.2, the process represented by Eq. A6.2 is nearly Gaussian, bounded, and has derivatives of all orders. From Eq. A6.2 we obtain by differentiation

$$\dot{x}(t) = -\sum_{k=1}^{N} a_k \omega_k \sin(\omega_k t + \phi_{0k}). \tag{A6.3}$$

In the limit of large N,

$$\sigma_x^2 = \frac{1}{2\pi}\int_0^\infty \Psi_x(\omega)\,d\omega, \tag{A6.4a}$$

$$\sigma_{\dot{x}}^2 = \frac{1}{2\pi}\int_0^\infty \Psi_{\dot{x}}(\omega)\,d\omega = [2/\pi]\int_0^\infty \omega^2\Psi_x(\omega)\,d\omega \tag{A6.4b}$$

(see Eq. 4.1.18). To calculate $\sigma_{x\dot{x}}$ we use the definition of the covariance (Eq. A5.12c) and Eqs. A6.2, A6.3, that is,

$$\sigma_{x\dot{x}} = -E\left[\sum_{j=1}^{N} a_j\cos(\omega_j t + \phi_{0j})\sum_{k=1}^{N} a_k\omega_k\sin(\omega_k t + \phi_{0k})\right]. \tag{A6.5}$$

Recalling that the random phase angles ϕ_{0i} $(i = 1, 2, \ldots, n)$ are independent and have probability density $1/(2\pi)$ in the interval $\{0, 2\pi\}$, it follows from the definition of the expectation (i.e., from the generalization to the multivariate case of Eq. A5.12a) that the covariance $\sigma_{x\dot{x}}$ vanishes. By virtue of Eq. A5.12d so does the correlation coefficient ρ. The probability density function of the time-dependent functions x, $\dot{x}$ is then given by Eq. A5.15 in which $\rho = 0$, that is,

$$f_{x\dot{x}}(x, \dot{x}; t) = \frac{1}{2\pi\sigma_x\sigma_{\dot{x}}}\exp\left[-\frac{1}{2}\left(\frac{x^2}{\sigma_x^2} + \frac{\dot{x}^2}{\sigma_{\dot{x}}^2}\right)\right]. \tag{A6.6}$$

(By virtue of the stationarity of the process $x(t)$, in Eq. A.6.6 we may write t instead of t_1.)

The probability that the process will go through the value k with a positive slope during the interval $t, t + dt$ is obtained immediately from Eq. A6.1 and Eq. A6.6 in which we substitute k for x. The result is

$$p_{k+dt} = \nu\exp(-\kappa^2/2)\,dt, \quad \nu = (1/2\pi)(\sigma_{\dot{x}}/\sigma_x), \quad \kappa = k/\sigma_x. \tag{A6.7a,b,c}$$

The expected number of upcrossings of the level k per unit time is called the upcrossing rate of the level k and is obtained by integrating Eq. A6.7a over an interval of one unit of time, that is,

$$\tau_u^{-1} = \nu\exp(-\kappa^2/2), \tag{A6.8}$$

where τ_u denotes the expected time between consecutive upcrossings of the level k. In particular, ν is the upcrossing rate of the level $k = 0$, that is, the zero upcrossing rate.

If the mean of the process is $-k$ instead of zero, Eq. A6.8 represents the zero upcrossing rate of the process $x(t)$.

Appendix A7

Mean Escape Rate τ_ϵ^{-1} for Systems Excited by White Noise

In the subsequent derivations we follow Soong and Grigoriu (1993). The system 5.5.1 excited by white noise is a two-dimensional vector process $\{x, \dot{x}\}$ for which the joint probability distribution may be obtained by using the Fokker-Planck equation (Risken, 1984; Soize, 1994). Once that distribution is obtained, the mean escape rate can be calculated by following steps similar to those that yielded the expression for the mean upcrossing rate (Appendix A6).

The expression for the steady-state Fokker-Planck equation corresponding to Eq. 5.5.1 is

$$-\frac{\partial}{\partial x}\left(\frac{\partial H}{\partial \dot{x}} f_{x,\dot{x}}\right) - \frac{\partial}{\partial \dot{x}}\left[\left(-\frac{\partial H}{\partial x} - \epsilon\beta\frac{\partial H}{\partial \dot{x}}\right) f_{x,\dot{x}}\right] + \frac{(\epsilon\gamma)^2}{2}\frac{\partial^2 f_{x,\dot{x}}}{\partial \dot{x}^2} = 0 \tag{A7.1}$$

(Soong and Grigoriu, 1993, p. 219), where $H = V(x) + \dot{x}^2/2$. We seek a solution of the form $f_{x,\dot{x}} = f_{x,\dot{x}}(H)$. Substituting that solution into Eq. A7.1 we obtain

$$\frac{\partial}{\partial \dot{x}}\left[\epsilon\beta\frac{\partial H}{\partial \dot{x}} f_{x,\dot{x}}(H)\right] + \frac{(\epsilon\gamma)^2}{2}\frac{\partial^2 f_{x,\dot{x}}(H)}{\partial \dot{x}^2} = 0. \tag{A7.2}$$

Integration of Eq. A7.2 yields

$$\epsilon\beta\frac{\partial H}{\partial \dot{x}} f_{x,\dot{x}}(H) + \frac{(\epsilon\gamma)^2}{2}\frac{\partial f_{x,\dot{x}}(H)}{\partial \dot{x}} = \phi(x). \tag{A7.3}$$

The left-hand side of Eq. A7.3 vanishes as $\dot{x} \to \infty$. Therefore we must have $\phi(x) \equiv 0$.

Since

$$\frac{\partial f_{x\dot{x}}(H)}{\partial \dot{x}} = \frac{df_{x,\dot{x}}(H)}{dH}\frac{\partial H}{\partial \dot{x}} \tag{A7.4}$$

and

$$\frac{\partial H}{\partial \dot{x}} \neq 0, \tag{A7.5}$$

we can write

$$\epsilon\beta f_{x,\dot{x}}(H) + \frac{(\epsilon\gamma)^2}{2}\frac{df_{x,\dot{x}}(H)}{dH} = 0. \tag{A7.6}$$

Therefore,

$$f_{x\dot{x}}(H) = \text{const} \times \exp\left(-\frac{2\epsilon\beta}{(\epsilon\gamma)^2}H\right), \tag{A7.7}$$

or

$$\begin{aligned} f_{x\dot{x}}(x, \dot{x}) &= \text{const} \times \exp[-2\epsilon\beta V(x)/(\epsilon\gamma)^2] \\ &\quad \times \exp[-2\epsilon\beta(\dot{x}^2/2)/(\epsilon\gamma)^2] \qquad \text{(A7.8a)} \\ &= a\ \exp[-2\epsilon\beta V(x)/(\epsilon\gamma)^2][2\pi(\epsilon\gamma)^2/(2\epsilon\beta)]^{-1/2} \\ &\quad \times \exp\{-\dot{x}^2/[2(\epsilon\gamma)^2/(2\epsilon\beta)]\} \qquad \text{(A7.8b)} \end{aligned}$$

where a is a normalization constant. Equation A7.8 shows that $x(t)$ and $\dot{x}(t)$ are independent at any time t (see Eq. A5.11), and that $\dot{x}$ is Gaussian with zero mean and variance $\sigma_{\dot{x}}^2 = (\epsilon\gamma)^2/(2\epsilon\beta)$ (see Eq. A5.14).

We now seek the mean upcrossing rate of the process $x(t)$, that is, the mean rate of occurrence of the event, denoted A, that the process crosses the line $x = b$ in an upward direction. The probability of occurrence of this event during a time interval dt in the neighborhood of the time t is

$$\nu_b\, dt = dt \int_0^\infty \dot{x} f_{x,\dot{x}}(b, \dot{x})\, d\dot{x}, \tag{A7.9}$$

where the mean upcrossing rate ν_b is the probability of occurrence of event A per unit time (Eq. A6.1). From Eq. A7.9,

$$\nu_b = f_x(b) \int_0^\infty \dot{x} f_{\dot{x}}(\dot{x})\, d\dot{x} = f_x(b)[\epsilon/(4\pi\beta)]^{1/2}\gamma. \tag{A7.10}$$

In Eq. A7.10 we used the fact that $x(t)$ and $\dot{x}(t)$ are independent and that $\dot{x}(t)$ is Gaussian with variance $\sigma_{\dot{x}}^2 = \epsilon\gamma^2/(2\beta)$. Also, it follows from Eq. A7.8b that

$$f_x(b) = a \exp\{-[2\beta/(\epsilon\gamma^2)]V(b)\}. \tag{A7.11}$$

The expression for ν_b follows from Eqs. A7.10 and A7.11. The normalization constant a is obtained from the condition that the area under the probability density function $f_x(x)$ is unity.

References

Allen, J. S., Samelson, R. M., and Newberger, P. A., 1991, "Chaos in a model of forced quasi-geostrophic flow over topography: an application of Melnikov's method," *Journal of Fluid Mechanics* **226** 511–547.

Argoul, F., Arneodo, A., and Richetti, P., 1991, "Dynamics in the Belousov-Zhabotinskii reaction: From Rössler's intuition to experimental evidence for Shilnikov's homoclinic chaos," in *A chaotic hierarchy*, Baier, G., and Klein, M., eds., World Scientific, Singapore.

Arrowsmith, D. K., and Place, C. M., 1990, *An introduction to dynamical systems,* Cambridge University Press, Cambridge.

Basios, V., Bountis, T., and Nicolis, G., 1999, "Controlling the onset of homoclinic chaos due to parametric noise," *Physics Letters A* **251** 250–258.

Beigie, D., Leonard, A., and Wiggins, S., 1991, "Chaotic transport in the homoclinic and heteroclinic tangle region of quasiperiodically forced two-dimensional dynamical systems," *Nonlinearity* **4** 775–819.

Benettin, G., Galgani, L., and Strelcyn, J. M., 1976, "Kolmogorov entropy and numerical experiments," *Physical Review A* **14** 2338–2345.

Bergé, P., Pomeau, Y., and Vidal, C., 1984, *Order within chaos,* Wiley, New York.

Bishop, S. R., and Thompson, J. M. T., 1999, "Stability of phase-locked loops," Centre for Nonlinear Dynamics and its Applications, University College, London.

Booker, S. M., and Smith, P. D., 1999, "Optimal modulations for forcing a PLL FM demodulator into chaotic behavior," Dept. of Mathematics, University of Dundee, Dundee, DD1 4HN, U.K.

Brunsden, V., Cortell, J., and Holmes, P., 1989, "Power spectra of chaotic vibrations of a buckled beam," *Journal of Sound and Vibration* **130** 1–25.

Charney, J. C., and DeVore, J. G., 1979, "Multiple flow equilibria in the atmosphere and blocking," *Journal of Atmospheric Sciences* **36** 1205–1216.

Chow, S. N., and Yamashita, M., 1992, "Geometry of the Melnikov vector," in *Nonlinear Equations in the Applied Sciences*, Academic, New York, pp. 79–148.

Cicogna, G., and Fronzoni, L., 1990, "Effects of parametric perturbations on the onset of chaos in the Josephson-junction model: Theory and analog experiments," *Physical Review A* **42** 1901–1906.

Cicogna G., and Fronzoni L., 1993, "Modifying the onset of homoclinic chaos: Application to a bistable potential," *Physical Review E* **47** 4585–4589.

Collins, J. J., Imhoff, T. T., and Grigg, P., 1996, "Noise-enhanced tactile sensation," *Nature* **383** 770–771.

Cook, G. R., and Simiu, E., 1991, "Periodic and chaotic oscillations of modified Stoker column," *Journal of Engineering Mechanics* **117** 2049–2064.

Encyclopedia of Statistical Sciences, 1985, Wiley, New York, Vol. 6.

Eckmann, J. P., and Ruelle, D., 1992, "Fundamental limitations for estimating dimensions and Lyapounov exponents in dynamical systems," *Physica D* **56** 185–187.

Falzarano, J. M., Shaw, S. W., and Troesch, A. W., 1992, "Application of global methods for analyzing dynamical systems to ship rolling motions and capsizing," *International Journal of Bifurcation and Chaos* **2** 101–116.

Feder, J., 1988, *Fractals,* Plenum, New York.

Fenichel, N., 1971, "Persistence and smoothness of invariant manifolds for flows," *Indiana University Mathematics Journal* **21** 193–225.

Franaszek, M., 1996, "Cutoff frequency of experimentally generated noise: A Melnikov approach," *Physical Review E* **54** 1–3.

Franaszek, M., and Simiu, E., 1995, "Crisis-induced intermittency and Melnikov scale factor," *Physics Letters A* **205** 137–142.

Franaszek, M., and Simiu, E, 1996a, "Stochastic resonance: A chaotic dynamics approach," *Physical Review E* **54** 1288–1304.

Franaszek, M., and Simiu, E., 1996b, "Noise-induced snap-through of buckled column with continuously distributed mass: A chaotic dynamics approach," *International Journal of Non-linear Mechanics* **31** 861–869.

Franaszek, M., and Simiu, E., 1998, "Auditory nerve fiber modeling: A stochastic Melnikov approach," *Physical Review E* **57** 5870–5876.

Frey, M., 1996, "A Wiener filter, state space flux-optimal control against escape from a potential well," *IEEE Transactions on Automatic Control* **41** 216–223.

Frey, M., and Simiu, E., 1993, "Noise-induced chaos and phase space flux," *Physica D* **63** 321–340.

Frey, M., and Simiu, E., 1995, "Noise-induced transitions to chaos," *Proceedings, NATO Advanced Workshop on Spatio-Temporal Patterns in Nonequilibrium Complex Systems,* Cladis, P., and Pally-Muhoray, P., eds., Santa Fe Institute in the Sciences of Complexity, Addison-Wesley, Reading, MA pp. 529–544.

Frey, M., and Simiu, E., 1996, "Phase space transport and control of escape from a potential well," *Physica D* **95**, 128–143.

Genchev, Z., Ivanov, Z., and Todorov, B., 1983, "Effect of periodic perturbation on radio frequency model of Josephson junction," *IEEE Transactions on Circuits and Systems* **CAS-30** 633–636.

Ghosh, S., Chang, H.-C., and Sen, M., 1992, "Heat-transfer enhancement due to slender recirculation and chaotic transport between counter-rotating eccentric cylinders," *Journal of Fluid Mechanics* **238** 119–154.

Grigoriu, M., 1995, *Applied non-Gaussian processes: Examples, theory, simulation, linear random vibrations, and MATLAB solutions*, Prentice-Hall, Englewood Cliffs, NJ.

Guckenheimer, J., and Holmes, P., 1986, *Nonlinear oscillations, dynamical systems, and bifurcations of vector fields,* Springer-Verlag, New York.

Gundlach, V. M., 1995, "Random homoclinic orbits," *Random and Computational Dynamics* **3** 1–33.

Hänggi, P., Jung, P., Zerbe, C., and Moss, F., 1993, "Can colored noise improve stochastic resonance?" *Journal of Statistical Physics,* **70** 25–47.

Hind, J. E., Anderson, D. J., Brugge, J. F., and Rose, J. R., 1967, "Coding of information pertaining to paired low-frequency tones in single auditory nerve fibers of the squirrel monkey," *Journal of Neurophysiology* **30** 794–816.

Hochmair-Desoyer, I. J., Hochmair, E. S., Motz, H., and Rattay, F., 1984, "A model for the electrostimulation of the Nervus acusticus," *Neuroscience* **13** 553–562.

Holmes, P., and Marsden, J., 1981, "A partial differential equation with infinitely many periodic orbits: Chaotic oscillations of a forced beam," *Archive for Rational Mechanics Analysis* **76** 135–166.

Hsieh, S. R., Troesch, A. W., and Shaw, S. W., 1994, "A nonlinear probabilistic method for predicting vessel capsizing in random beam seas," *Proceedings of the Royal Society of London A* **446** 195–211.

Kac, M., 1959. *Statistical Independence in Probability Analysis and Number Theory*, Wiley, New York.

Kautz, R. L., 1985, "Chaos and thermal noise in the rf-biased Josephson junction," *Journal of Applied Physics* **58** 424–440.

Larson, H. J., and Shubert, B. O., 1979, *Probabilistic models in engineering sciences,* Wiley, New York.

Lichtenberg, A. J., and Lieberman, M. A., 1992, *Regular and chaotic dynamics,* 2nd ed. Springer-Verlag, New York.

Lima, R., and Pettini, M., 1990, "Suppression of chaos by resonant parametric perturbations," *Physical Review A* **41** 726–733.

Lindenberg, K., West, B. J., and Masoliver, J., 1989, "First passage time problems for non-Markovian processes," in *Noise in nonlinear dynamical systems,* Moss, F., and McClintock, P. V. E., eds., Cambridge University Press, Cambridge, Vol. 1, pp. 110–160.

Longtin, A., 1993, "Stochastic resonance in neuron models," *Journal of Statistical Physics* **70** 309–327.

Lorenz, E. N., 1963, "Deterministic non-periodic flow," *Journal of Atmospheric Sciences* **20** 130–141.

MacKay, R. S., Meiss, J. D., and Percival, I. C., 1984, "Transport in Hamiltonian systems," *Physica D* **13** 55–81.

Mandelbrot, B. B., 1986, "Self-affine fractal sets," in *Fractals in physics,* Pietronero, L., and Tosati, E., eds., North-Holland, Amsterdam, pp. 3–28.

McNamara, B., and Wiesenfeld, K., 1989, "Theory of stochastic resonance," *Physical Review A* **39** 4854–4869.

Meirovich, L., *Analytical methods in vibrations,* Macmillan Collier-Macmillan, Toronto, 1967.

Melnikov, V. K., 1963, "On the stability of the center for time periodic perturbations," *Transactions of the Moscow Mathematical Association* **12** 1–57.

Meyer, K. R., and Sell, G. R., 1989, "Melnikov transforms, Bernoulli bundles, and almost periodic perturbations," *Transactions of the American Mathematical Society* **314** 63–105.

Moon, F., 1992, *Chaotic vibrations*, Wiley-Interscience, New York.

Moon, F., and Holmes, P., 1979, "A magnetoelastic strange attractor," *Journal of Sound and Vibration* **65** 275–296.

Moon, F., and Li, G.-X., 1985, "Fractal basin boundaries and homoclinic orbits for periodic motion in a two-well potential," *Physica D* **17** 99–108.

Moss, F., Pierson, D., and O'Gorman, D., 1994, "Stochastic resonance: Tutorial and update," *International Journal of Bifurcation and Chaos,* **4** 1383–1397.

National Safety Transportation Board, 1979, "Grounding and capsizing of the clam dredger *Patti-B,*" NSTB Marine Accident Report.

NOAA, 1977, *Local Climatological Annual Summaries,* National Oceanic and Atmospheric Administration, Environmental Data Service, Asheville, NC.

Ott, E., 1997, *Chaos in dynamical systems*, Cambridge University Press, Cambridge.

Papoulis, A., 1962, *The Fourier transform and its applications,* McGraw-Hill, New York.

Parkinson, G. V., and Smith, J. D., 1964, "The Square Prism as an Aeroelastic Nonlinear Oscillator," *Quarterly Journal of Mechanics and Applied Mathematics* **17** 225–239.

Poincaré, H., 1892, *Les méthodes nouvelles de la mécanique céleste,* Gauthier-Villars, Paris.

Rice, S. O., 1954, "Mathematical analysis of random noise," in *Selected papers in noise and stochastic processes,* Wax, A., ed., Dover, New York.

Risken, H., 1984, *The Fokker-Planck equation: Methods of solution and applications,* Springer-Verlag, New York.

Rose, J. R., Brugge, J. F., Anderson, D., and Hind, J. E., 1967, "Phase-locked response to low-frequency tones in single auditory nerve fibers of the squirrel monkey," *Journal of Neurophysiology* **30** 769–783.

Ruggero, M. A., 1973, "Response to noise of auditory nerve fibers in the squirrel monkey," *Journal of Neurophysiology* **36** 569–587.

Sanders, J. A., 1982, "Melnikov method and averaging," *Celestial Mechanics* **28** 171–181.

Shinozuka, M., 1971, "Simulation of multivariate and multidimensional random processes," *Journal of Acoustical Society of America* **49** 347–357.

Shulgin, B., Neiman, A., and Anishchenko, V., 1995, "Mean switching frequency locking in stochastic bistable systems driven by a periodic force," *Physical Review Letters* **75** 4157–4159.

Simiu, E., 1996, "Melnikov process for stochastically perturbed slowly varying oscillators: Application to a model of wind-driven coastal currents," *Journal of Applied Mechanics* **63** 429–436.

Simiu, E., and Cook, G. R., 1991, "Chaotic motions of self-excited forced and autonomous square prisms," *Journal of Engineering Mechanics* **117** 241–259.

Simiu, E., and Cook, G. R., 1992, "Empirical fluid-elastic models and chaotic galloping: A case study, *Journal of Sound and Vibration* **154** 45–66.

Simiu, E., and Franaszek, M., 1997, "Melnikov-based open-loop control of escape for a class of nonlinear systems," *ASME Journal of Dynamical Systems, Measurement, and Control* **119** 590–594.

Simiu, E., and Frey, M., 1993a, "Melnikov function and homoclinic chaos induced by weak perturbations," *Physical Review E* **48** 3190–3192.

Simiu, E., and Frey, M., 1993b, "Spectrum of the stochastically forced Duffing-Holmes oscillator," *Physics Letters A* **177** 199–202.

Simiu, E., and Frey, M., 1996a, "Noise-induced sensitivity to initial conditions," in *Proceedings of the workshop on fluctuations and order: The new synthesis,* Millonas, M., ed., Springer-Verlag, New York, pp. 81–90.

Simiu, E., Frey, M., and Grigoriu, M., 1991, "Necessary conditions for homoclinic chaos induced by additive noise," in *Computational stochastic mechanics,* Spanos, P., and Brebbia, C., eds., Elsevier, New York.

Simiu, E., and Grigoriu, M., 1995, "Non-Gaussian noise effects on reliability of multistable systems, *Journal of Offshore Mechanics and Arctic Engineering* **117** 166–170.

Simiu, E., and Scanlan, R. H., 1996, *Wind effects on structures,* 3rd ed., Wiley-Interscience, New York.

Sivathanu, Y., Hagwood, C., and Simiu, E., 1995, "Exits in multistable systems excited by coin-toss square wave dichotomous noise: A chaotic dynamics approach," *Physical Review E* **52** 4669–4675.

Soize, C., 1994, *The Fokker-Planck equation for stochastic dynamical systems and its explicit steady state solutions,* World Scientific, Singapore.

Soong, T. T., and Grigoriu, M., 1993, *Random vibrations of mechanical and structural systems,* Prentice-Hall, Englewood Cliffs, NJ.

Steinkamp, O., 1999, *Melnikov's method and homoclinic chaos for random dynamical systems,* Dissertation zur Erlangung des Grades eines Doktors der Naturwissenschaften, vorgelegt im Fachbereich Mathematik und Informatik der Universität Bremen, Logos Verlag, Berlin.

Stoker, J. J., 1950, *Nonlinear vibrations*, Interscience, New York.

Stone, E., and Holmes, P., 1990, "Random perturbations of heteroclinic attractors," *SIAM Journal on Applied Mathematics* **50** 726–745.

Tan, N., and Radmore, P., 1995, "Alternative approaches to Melnikov analysis for forced oscillators," *Journal of Sound and Vibration* **187** 815–824.

Thompson, J. M. T., and Stewart, B., 1986, *Nonlinear dynamics and chaos*, Wiley, New York.

Tseng, W. Y., and Dugundji, J., 1971, "Nonlinear vibrations of a buckled beam under harmonic excitations," *Journal of Applied Mechanics* **38** 467–476

Vaicaitis, R., and Simiu, E., 1977, "Nonlinear pressure terms and along-wind response," *Journal of the Structural Division, ASCE* **103** 903–906.

Van den Broeck, C., and Nicolis, C., 1993, "Noise-induced sensitivity to initial conditions in stochastic dynamical systems," *Physical Review E* **48** 4845–4846.

van der Hoven, I., 1957, "Power spectrum of horizontal wind speed in the frequency range from 0.0007 to 900 cycles per hour," *Journal of Meteorology* **14** 160–163.

Verhulst, F., 1990, *Nonlinear differential equations and dynamical systems*, Springer-Verlag, New York.

von Mises, R., 1957, *Probability, statistics, and truth*, George Unwin and Allen, London/MacMillan, New York.

Whalen, T. M., 1996, "Stability of multi-degree of freedom stochastic nonlinear systems," in *Nonlinear dynamics and controls,* Bajab, H. et al., eds., ASME International Mechanical Engineering Conference, American Society of Mechanical Engineers, New York.

Wiggins, S., 1988, *Global bifurcations and chaos: Analytical methods*, Springer-Verlag, New York.

Wiggins, S., 1990, *Introduction to applied nonlinear dynamical systems and chaos,* Springer-Verlag, New York.

Wiggins, S., 1992, *Chaotic transport in dynamical systems*, Springer-Verlag, New York.

Wiggins, S., and Holmes, P., 1987, 1988, "Homoclinic orbits in slowly varying oscillators," *SIAM Journal on Mathematical Analysis* **18** 612–629; Erratum: **19** 1254–1255.

Wiggins, S., and Shaw, S. W., 1988, "Chaos and three-dimensional horseshoes in slowly varying oscillators," *Journal of Applied Mechanics* **55** 959–968.

Index

College Library